Multivariate Data Integration Using R

Computational Biology Series

About the Series:

This series aims to capture new developments in computational biology, as well as high-quality work summarizing or contributing to more established topics. Publishing a broad range of reference works, textbooks, and handbooks, the series is designed to appeal to students, researchers, and professionals in all areas of computational biology, including genomics, proteomics, and cancer computational biology, as well as interdisciplinary researchers involved in associated fields, such as bioinformatics and systems biology.

Introduction to Bioinformatics with R: A Practical Guide for Biologists
Edward Curry

Analyzing High-Dimensional Gene Expression and DNA Methylation Data with R
Hongmei Zhang

Introduction to Computational Proteomics
Golan Yona

Glycome Informatics: Methods and Applications
Kiyoko F. Aoki-Kinoshita

Computational Biology: A Statistical Mechanics Perspective
Ralf Blossey

Computational Hydrodynamics of Capsules and Biological Cells
Constantine Pozrikidis

Computational Systems Biology Approaches in Cancer Research
Inna Kuperstein, Emmanuel Barillot

Clustering in Bioinformatics and Drug Discovery
John David MacCuish, Norah E. MacCuish

Metabolomics: Practical Guide to Design and Analysis
Ron Wehrens, Reza Salek

An Introduction to Systems Biology: Design Principles of Biological Circuits
2nd Edition
Uri Alon

Computational Biology: A Statistical Mechanics Perspective
Second Edition
Ralf Blossey

Stochastic Modelling for Systems Biology
Third Edition
Darren J. Wilkinson

Computational Genomics with R
Altuna Akalin, Bora Uyar, Vedran Franke, Jonathan Ronen

An Introduction to Computational Systems Biology: Systems-level Modelling of Cellular Networks
Karthik Raman

Virus Bioinformatics
Dmitrij Frishman, Manuela Marz

Multivariate Data Integration Using R: Methods and Applications with the mixOmics Package
Kim-Anh LeCao, Zoe Marie Welham

Bioinformatics
A Practical Guide to NCBI Databases and Sequence Alignments
Hamid D. Ismail

For more information about this series please visit:
https://www.routledge.com/Chapman--HallCRC-Computational-Biology-Series/book-series/CRCCBS

Multivariate Data Integration Using R

Methods and Applications with the mixOmics Package

Kim-Anh Lê Cao
Zoe Welham

CRC Press
Taylor & Francis Group
Boca Raton London New York

CRC Press is an imprint of the
Taylor & Francis Group, an **informa** business

A CHAPMAN & HALL BOOK

First edition published 2022
by CRC Press
6000 Broken Sound Parkway NW, Suite 300, Boca Raton, FL 33487-2742

and by CRC Press
2 Park Square, Milton Park, Abingdon, Oxon, OX14 4RN

ISBN: 9780367460945 (hbk)
ISBN: 9781032128078 (pbk)
ISBN: 9781003026860 (ebk)

DOI: 10.1201/9781003026860

This book has been prepared from camera-ready copy provided by the authors.

From Kim-Anh Lê Cao:
To my parents, Betty and Huy Lê Cao
To the mixOmics team and our mixOmics users,
And to my co-author Zoe Welham without whom this book would not have existed.

For Zoe Welham:
To Joy and Tamara Welham, for their continual support

Contents

Preface

Context and scope

Modern high-throughput technologies generate information about thousands of biological molecules at different cellular levels in a biological system, leading to several types of omics studies (e.g. transcriptomics as the study of messenger RNA molecules expressed from the genes of an organism, or proteomics as the study of proteins expressed by a cell, tissue, or organism). However, a reductionist approach that considers each of these molecules individually does not fully describe an organism in its environment. Rather, we use multivariate analysis to investigate the simultaneous and complex relationships that occur in molecular pathways. In addition, to obtain a holistic picture of a complete biological system, we propose to integrate multiple layers of information using recent computational tools we have developed through the mixOmics project.

mixOmics is an international endeavour that encompasses methodological developments, software implementation, and applications to biological and biomedical problems to address some of the challenges of omics data integration. We have trained students and researchers in essential statistical and data analysis skills via our numerous multi-day workshops to build capacity in best practice statistical analysis and advance the field of computational statistics for biology. The goal of this book is to provide guidance in applying multivariate dimension reduction techniques for the integration of high-throughput biological data, allowing readers to obtain new and deeper insights into biological mechanisms and biomedical problems.

Who is this book for?

This book is suitable for biologists, computational biologists, and bioinformaticians who generate and work with high-throughput omics data. Such data include – but are not restricted to – transcriptomics, epigenomics, proteomics, metabolomics, the microbiome, and clinical data. Our book is dedicated to research postgraduate students and scientists at any career stage, and can be used for teaching specialised multi-disciplinary undergraduate and Masters's courses. Data analysts with a basic level of R programming will benefit most from this resource. The book is organised into three distinct parts, where each part can be skimmed according to the level and interest of the reader. Each chapter contains different levels of information, and the most technical chapters can be skipped during a first read.

Overview of methods in `mixOmics`

The `mixOmics` package focuses on multivariate analysis which examines more than two variables simultaneously to integrate different types of variables (e.g. genes, proteins, metabolites). We use dimension reduction techniques applicable to a wide range of data analysis types. Our analyses can be descriptive, exploratory, or focus on modeling or prediction. Our

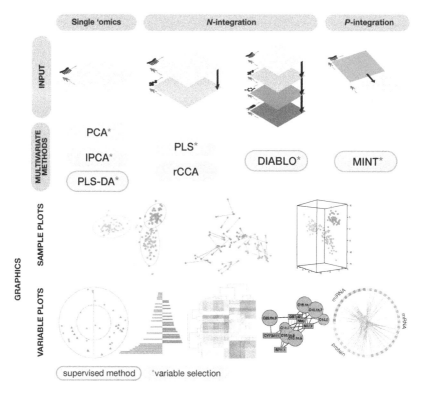

FIGURE 1: *Overview of the methods implemented in the* `mixOmics` *package for the exploration and integration of multiple data sets.* This book aims to guide the data analyst in constructing the research question, applying the appropriate multivariate techniques, and interpreting the resulting graphics.

aim is to summarise these large biological data sets to elucidate similarities between samples, between variables, and the relationship between samples and variables. The `mixOmics` package provides a range of methods to answer different kinds of biological questions, for example to:

- Highlight patterns pertaining to the major sources of variation in the data (e.g. Principal Component Analysis),
- Segregate samples according to their known group and predict group membership of new samples (e.g. Partial Least Squares Discriminant Analysis),
- Identify agreement between multiple data sets (e.g. Canonical Correlation Analysis, Partial Least Squares regression, and other variants),
- Identify molecular signatures across multiple data sets with sparse methods that achieve variable selection.

Key methodological concepts in `mixOmics`

Methods in `mixOmics` are based on *matrix factorisation* techniques, which offer great flexibility in analysing and integrating multiple data sets in a holistic manner. We use *dimension reduction* combined with *feature selection* to summarise the main characteristics of the data and posit novel biological hypotheses.

Dimension reduction is achieved by combining all original variables into a smaller number of artificial *components* that summarise patterns in the original data.

The `mixOmics` package is unique in providing novel multivariate techniques that enable feature selection to identify *molecular signatures*. Feature selection refers to identifying variables that best explain, or predict, the outcome variable (e.g. group membership, or disease status) of interest. Variables deemed irrelevant according to the specific statistical criterion we use in the methods are not taken into account when calculating the components.

Data integration methods use *data projection* techniques to maximise the covariance, or the correlation between, omics data sets. We propose two types of data integration, whether on the same N samples, or on the same P variables (Figure 1).

Finally, our methods can provide either *unsupervised* or *supervised* analyses. Unsupervised analyses are exploratory: any information about sample group membership, or outcome, is disregarded, and data are explored based on their *variance* or *correlation* structure. Supervised analyses aim to segregate sample groups known *a priori* (e.g. disease status, treatments) and identify variables (i.e. biomarker candidates, or molecular signatures) that either explain or separate sample groups.

These concepts will be explained further in Part I.

To aid in interpreting analysis results, `mixOmics` provides insightful graphical plots designed to highlight patterns in both the sample and variable dimensions uncovered by each method (Figure 1).

Concepts not covered

Each `mixOmics` method corresponds to an underlying statistical model. However, the methods we present radically differ from univariate formulations as they do not test one variable at a time, or produce *p*-values. In that sense, multivariate methods can be considered *exploratory* as they do not enable statistical inference. Our wide range of methods come in many different flavours and can be applied also for predictive purposes, as we detail in this book. 'Classical' univariate statistical inference methods can still be used in our analysis framework after the identification of molecular signatures, as our methods aim to generate novel biological hypotheses.

Who is 'mixOmics'?

The mixOmics project has been developed between France, Australia and Canada since 2009, when the first version of the package was submitted to the CRAN[1]. Our team is composed of core members from the University of Melbourne, Australia, and the Université de Toulouse, France. The team also includes several key contributors and collaborators.

The package implements more than nineteen multivariate and sparse methodologies for omics data exploration, integration, and biomarker discovery for different biological settings, amongst which thirteen were developed by our team (see our list of publications in Section 14.8). Originally, all methods were designed for omics data, however, their application is not limited to biological data only. Other applications where integration is required can be considered, but mostly for cases where the predictor variables are continuous.

---

[1]The Comprehensive R Architecture Network https://www.cran.r-project.org

The package is currently available from Bioconductor[2], with a development version available on GitHub[3]. We continue to maintain and improve the package via new methods, code optimisation and efficient memory storage of R objects.

About this book

Part I: Modern biology and multivariate analysis introduces fundamental concepts in multivariate analysis. *Multi-omics and biological systems* (Chapter 1) compares and contrasts multivariate and univariate analysis, and outlines the advantages and challenges of multivariate analyses. *The Cycle of Analysis* (Chapter 2) details the necessary steps in planning, designing and conducting multivariate analyses. *Key multivariate concepts and dimension reduction in `mixOmics`* (Chapter 3) describes measures of dispersion and association, and introduces key methods in `mixOmics` to manage large data, such as dimension reduction using matrix factorisation and feature selection. *Choose the right method for the right question in `mixOmics`* (Chapter 4) provides an overview of the methods available in `mixOmics` and the types of biological questions these methods can answer.

Part II: mixOmics under the hood provides a deeper understanding of the statistical concepts underlying the methods presented in Part III. *Projection to Latent Structures (PLS)* (Chapter 5) illustrates the different types of algorithms used to solve Principal Component Analysis. We detail in particular the iterative PLS algorithm that projects data onto latent structures (components) for matrix decomposition and dimension reduction, as this algorithm forms the basis of most of our methods. *Visualisation for data integration* (Chapter 6) showcases the variety of graphical outputs offered in `mixOmics` to complement each method. *Performance assessment in supervised analyses* (Chapter 7) describes the techniques employed to evaluate the results of the analyses.

Part III: mixOmics in action provides detailed case studies that apply each method in `mixOmics` to answer pertinent biological questions, complete with example R code and insightful plots. We begin with *mixOmics: get started* (Chapter 8) to guide the novice analyst in using the R platform for data analysis. Each subsequent chapter is dedicated to one method implemented in `mixOmics`. In *Principal Component Analysis (PCA)* (Chapter 9) and *PLS - Discriminant Analysis (PLS-DA)* (Chapter 12), we introduce different multivariate methods for single omics analysis. The N-integration framework is introduced in *PLS* (Chapter 10) and *Canonical Correlation Analysis* (Chapter 11) for two omics, and $N-data\ integration$ (DIABLO, Chapter 13) for multi-omics integration. $P-data\ integration$ (MINT, Chapter 14) introduces our latest developments for P-integration to combine independent omics studies. Each of these chapters is organised as follows:

- Aim of the method,
- Research question framed biologically and statistically,
- Principles of the method,
- Input arguments and key outputs,
- Introduction of the case study,
- Quick start R command lines,
- Further options to go deeper into the analysis,
- Frequently Asked Questions,
- Technical methodological details in each Appendix.

---

[2]https://www.bioconductor.org/packages/release/bioc/html/mixOmics.html
[3]https://github.com/mixOmicsTeam/

Additional resources related to this book

In addition to the R package, the mixOmics project includes a website with extensive tutorials in `http://www.mixOmics.org`. The R code of each chapter is also available on the website. Our readers can also register for our newsletter mailing list, and be part of the mixOmics community on GitHub and via our discussion forum `https://mixomics-users.discourse.group/`.

Authors

Dr Kim-Anh Lê Cao develops novel methods, software and tools to interpret big biological data and answer research questions efficiently. She is committed to statistical education to instill best analytical practice and has taught numerous statistical workshops for biologists and leads collaborative projects in medicine, fundamental biology or microbiology disciplines. Dr Kim-Anh Lê Cao has a mathematical engineering background and graduated with a PhD in Statistics from the Université de Toulouse, France. She is currently an Associate Professor in Statistical Genomics at the University of Melbourne. In 2019, Kim-Anh received the Australian Academy of Science's Moran Medal for her contributions to Applied Statistics in multidisciplinary collaborations. She has contributed to a leadership program for women in STEMM, including the international Homeward Bound which culminated in a trip to Antarctica, and Superstars of STEM from Science Technology Australia.

Zoe Welham completed a BSc in molecular biology and during this time developed an interest in the analysis of big data. She completed a Master of Bioinformatics with a focus on the statistical integration of different omics data in bowel cancer. She is currently a PhD candidate at the Kolling Institute in Sydney where she is furthering her research into bowel cancer with a focus on integrating microbiome data with other omics to characterise early bowel polyps. Her research interests include bioinformatics and biostatistics for many areas of biology and making that information accessible to the general public.

Part I

Modern biology and multivariate analysis

1

Multi-omics and biological systems

Technological advances such as next-generation sequencing and mass spectrometry generate a wealth of diverse biological information, allowing for the monitoring of thousands of variables, or dimensions, that describe a given sample or individual, hence the term 'high dimensional data'. Multi-omics variables represent molecules from different functional levels: for example, transcriptomics for the study of transcripts, proteomics for proteins, and metabolomics for metabolites. However, their complex nature requires an integrative, multidisciplinary approach to analysis that is not yet fully established.

Historically, the scientific community has adopted a *reductionist* approach to data analysis by characterising a very small number of genes or proteins in one experiment to assess specific hypotheses. A holistic approach allows for a deeper understanding of biological systems by adding two new facets to analysis (Figure 1.1): Firstly, by integrating data from different omic functional levels, we move from clarifying a linear process (e.g. the dysregulation of one or two genes) towards understanding the development, health, and disease of an ever-changing, dynamic, hierarchical *system*. Secondly, by adopting a hypotheses-free, data-driven approach, we can build integrated and coherent models to address novel, systems-level hypotheses that can be further validated through more traditional hypotheses.

1.1 Statistical approaches for reductionist or holistic analyses

Compared to a traditional reductionist analysis, multivariate multi-omics analysis drastically differs in its viewpoint and aims. We briefly introduce three types of analysis to illustrate this point:

A **univariate analysis** is a fundamentally reductionist, *hypothesis-driven* approach that is related to inferential statistics (introduced in Section 2.4). A hypothesis test is conducted on one variable (e.g. gene expression or protein abundance) independently from the other variables. Univariate methods make inferences about the population and measure the certainty of this inference through test statistics and p-values. Linear models, t-tests, F-tests, or non-parametric tests fit into a univariate analysis framework. Although interactions between variables are not considered in univariate analyses, when one variable is manipulated in a controlled experiment, we can often attribute the result to that particular variable. In omics studies, where multiple variables are monitored simultaneously, it is difficult to determine which variables influence the biology of interest.

A **bivariate analysis** considers two variables simultaneously, for example, to assess the association between the expression levels of two genes via correlation or linear regression. Such an analysis is often supported by visualisation through scatterplots but can quickly

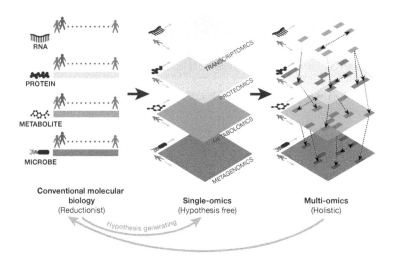

FIGURE 1.1: From reductionism to holism. Until recently, only a few molecules of a given omics type were analysed and related to other omics. The advent of high-throughput biology has ushered in an era of hypothesis-free approaches within a single type of omics data, and across multiple omics from the same set of samples. A holistic approach is now required to understand the different omic functional layers in a biological system and posit novel hypotheses that can be further validated with a traditional reductionist approach. We have omitted DNA as this data type needs to be handled differently in `mixOmics`, see Section 4.2.3.

become cumbersome when dealing with thousands of variables that are considered in a pairwise manner.

A **multivariate analysis** examines more than two variables simultaneously and potentially thousands at a time. In omics studies, this approach can lead to computational issues and inaccurate results, especially when the number of samples is much smaller than the number of variables. Several computational and statistical techniques have been revisited or developed for high-dimensional data. This book focuses on multivariate analyses and extends this to include the integration of multi-omics data sets.

1.2   Multi-omics and multivariate analyses

The aim of omics data integration is to identify associations and patterns amongst different variables and across different omics collected from a sample. Provided appropriate data analysis is conducted, the integration of multiple data sources may also consolidate our confidence in the results when consensus is observed from different experiments.

1.2.1   More than a 'scale up' of univariate analyses

The fundamental difference between multivariate and univariate analysis lies in the scope of the results obtained. Multivariate analysis can unravel groups of variables that share similar patterns in expression across different phenotypes, thus complementing each other to describe an outcome. A univariate analysis may declare the same variables as non-significant, as a variable's ability to explain the phenotype may be subtle and can be masked by individual variation, or confounders (Saccenti et al., 2014). However, with sufficiently powered data, univariate and multivariate methods are complementary and can help make sense of the data. For example, several multivariate and exploratory methods presented in this book can suggest promising candidate variables that can be further validated through experiments, reductionist approaches, and inferential statistics.

1.2.2   More than a fishing expedition

Multivariate analyses, which examine up to thousands of variables simultaneously, are often considered to be 'fishing expeditions'. This somewhat pejorative term refers to either conducting analyses without first specifying a testable hypothesis based on prior research, or, conducting several different analyses on the same data to 'fish' for a significant result regardless of its domain relevance. Indeed, examining a large number of variables can lead to statistically significant results purely by chance.

However, the integration of multi-omics data, with an appropriate experimental design set in an exploratory, rather than predictive approach, offers a tremendous opportunity for discovering associations between omics molecules (whether genes, transcripts, proteins, or metabolites), in normal, temporal or spatial changes, or in disease states. For example, one of our studies identified pathways that were never previously identified as relevant to ontogeny during the first week of human life (Lee et al., 2019). Multi-omics data integration has deepened our understanding of gene regulatory networks by including information from related molecules prior to validation of gene associations with a functional approach (Gligorijević and Pržulj, 2015) and has also efficiently improved functional annotations to proteins instead of using expensive and time-consuming experimental techniques (Ma et al., 2013). Multi-omics can also more easily characterise the relatively small number of genes associated with a particular disease by integrating multiple sources of information (Žitnik et al., 2013). Finally, it has further developed precision medicine by integrating patient- and disease-specific information with the aim to improve prognosis and clinical outcomes (Ritchie et al., 2015).

1.3   Shifting the analysis paradigm

Despite the potential advantages of high-dimensional data, we should keep in mind that quantity does not equal quality. Multivariate data integration is not straightforward: the analyses cannot be reduced to a mere concatenation of variables of different types, or by overlapping information between single data sets, as we illustrate in Figure 1.2. As such, we must shift our traditional view of analysing data.

Biological experimentation often employs univariate statistics to answer clear hypotheses

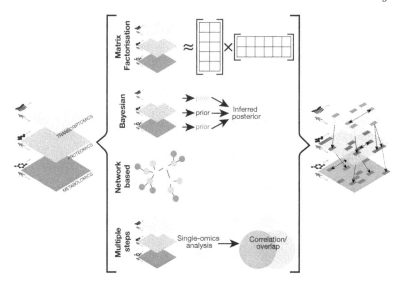

FIGURE 1.2: Types of methods for data integration. Methods for multi-omics data integration are still in active development, and can be broadly categorised into matrix factorisation techniques (the focus of this book), Bayesian, network-based, and multiple-step approaches. The latter deviates from data integration as it considers each data set individually before combining the results.

about the potential causal effect of a given molecule of interest. In high-dimensional data sets, this reductionist approach may not hold due to the sheer amount of molecules that are monitored, and their interactions that might be of biological interest. Therefore, exploratory, *data-driven* approaches are needed to extract information from noise and generate new hypotheses and knowledge. However, the lack of a clear, causal-driven hypothesis presents a challenging new paradigm in statistical analyses.

In univariate hypothesis testing, we report p-values to determine the significance of a statistical test conducted on a *single* variable. In a multivariate setting, however, a p-value assesses the statistical significance of a result while taking into account *all* variables simultaneously. In such analyses, permutation-based tests are common to assess how far from random a result is when the data are reshuffled, but other inference-based methods are currently being developed in the field of multivariate analysis (Wang and Xu, 2021). In `mixOmics` we do not offer such tests, but related methods propose permutation approaches to choose the parameters in the method (see Section 10.7.5).

1.4 Challenges with high-throughput data

There are multiple challenges associated with managing large amounts of biological data, pertaining to specific types of data as well as statistical analysis. To make reliable, valid, and meaningful interpretations, these challenges must be considered, ideally before data collection.

1.4.1 Overfitting

Multivariate omics analysis assesses many molecules that individually, or in combination, can explain the biological outcome of interest. However, these associations may be spurious, as the large number of features can often be combined in different ways to explain the outcome well, despite having no biological relevance. Overfitting occurs when a statistical model captures the noise along with the underlying pattern in the data: if we apply the same statistical model fitted on a high-dimensional data set to a similar but external study, we might obtain different results.[1] The problem of overfitting is a well-known issue in high-throughput biology (Hawkins, 2004). We can assess the amount of overfit using cross-validation or subsampling of the data, as described in Chapter 7.

1.4.2 Multi-collinearity and ill-posed problems

As the number of variables increase, the number of pairwise correlations also increases. *Multi-collinearity* poses a problem in most statistical analyses as these variables bring redundant and noisy information that decreases the precision of the statistical model. Correlations in high-throughput data sets are often spurious, especially when the number of biological samples, or individuals N, is small compared to the number of variables P[2]. The 'small N large P' problem is defined as *ill-posed*, as standard statistical inference methods assume N is much greater than P to generalise the results to the population the sample was drawn from. Ill-posed problems also lead to inaccurate computations.

1.4.3 Zero values and missing values

Data sets may contain a large number of zeros, depending on the type of omics studied and the platform that is used. This is particularly the case for microbiome, proteomics, and metabolomics data: a large number of zeros results in zero-inflated (skewed) data, which can impair methods that assume a normal distribution of the data. *Structural zeros*, or true zeros, reflect a true absence of the variable in the biological environment while *sampling zeros*, or false zeros, may not reflect reality due to experimental error, technological reasons, or an insufficient sample size (Blasco-Moreno et al., 2019). The challenge is whether to consider these zeros as a true zero or missing (coded as NA in R).

Methods that can handle missing values often assume they are 'missing at random', i.e. missingness is not related to a specific sample, individual, molecule, or type of omics platform. Some methods can estimate missing values, as we present in Appendix 9.A.

---

[1]Statistical models that overfit have low bias and high variance, meaning that they tend to be complex to fit the training data well, but do not predict well on test data (more details about the bias-variance tradeoff can be found in Friedman et al. (2001) Chapter 2).

[2]In our context, N can also refer to the number of cells in single cell assays, as we briefly mention in Section 14.6.

1.5   Challenges with multi-omics integration

Examining data holistically may lead to better biological understanding, but integrating multiple omics data sets is not a trivial task and raises another series of challenges.

1.5.1   Data heterogeneity

Different omics rely on different laboratory techniques and data extraction platforms, resulting in data sets of different formats, complexity, dimensionalities, information content, and scale, and may be processed using different bioinformatics tools. Therefore, data heterogeneity arises from biological *and* technical reasons and is the main analytical challenge to overcome.

1.5.2   Data size

Integrating multiple omics results in a drastic increase in the number of variables. A filtering step is often applied to remove irrelevant and noisy variables (see Section 8.1). However, the number of variables P still remains extremely large compared to the number of samples N, which raises computational as well as analytical issues.

1.5.3   Platforms

The data integration field is constantly evolving due to ever-advancing technologies with new platforms and protocols, each containing inherent technical biases and analytical challenges. It is crucial that data analysts swiftly adapt their analysis framework to keep apace with these omics-era demands. For example, single cell techniques are rapidly advancing, as are new protocols for their multi-omics analysis.

1.5.4   Expectations for analysis

The field of data integration has no set definition. Data integration can be managed biologically, bioinformatically, statistically, or at the interpretation steps (i.e. by overlapping biological interpretation once the statistical results are obtained). Therefore, the expectations for data integration are diverse; from exploration, and from a low to high-level understanding of the different omics data types. Despite recent advances in single cell sequencing, current technologies are still limited in their ability to parse omics interactions at precise functional levels. Thus, our expectations for data integration are limited, not only by the statistical methods but also by the technologies available to us.

1.5.5   Variety of analytical frameworks

Integrative techniques fully suited to multi-omics biological data are still in development and continue to expand[3]. Different types of techniques can be considered and broadly categorised into (Huang et al. (2017), Figure 1.2):

- Matrix factorisation techniques, where large data sets are decomposed into smaller sub-matrices to summarise information. These techniques use algebra and analysis to optimise specific statistical criteria and integrate different levels of information. Methods in `mixOmics` fit into this category and will be detailed in Chapter 3 and subsequent chapters,

- Bayesian methods, which use assumptions of prior distributions for each omics type to find correlations between data layers and infer posterior distributions,

- Network-based approaches, which use visual and symbolic representations of biological systems, with nodes representing molecules and edges as correlations between molecules, if they exist. Network-based methods are mostly applied for detecting significant genes within pathways, discovering sub-clusters, or finding co-expression network modules,

- Multiple-step approaches that first analyse each single omics data set individually before combining the results based on their overlap (e.g. at the gene level of a molecular signature) or correlation. This type of approach technically deviates from data integration but is commonly used.

1.6   Summary

Modern biological data are high dimensional; they include up to thousands of molecular entities (e.g. genes, proteins, or epigenetic markers) per sample. Integrating these rich data sets can potentially uncover the hierarchical and holistic mechanisms that govern biological pathways. While classical, reductionist, univariate methods ignore these molecular interactions, multivariate, integrative methods offer a promising alternative to obtain a more complete picture of a biological system. Thus, univariate and multivariate methods are different approaches with very little overlap in results but have the advantage of complementarity.

The advent of high-throughput technology has revealed a complex world of multi-omics molecular systems that can be unraveled with appropriate integration methods. However, multivariate methods able to manage high-dimensional and multi-omics data are yet to be fully developed. The methods presented in this book mitigate some of these challenges and will help to reveal patterns in omics data, thus forging new insights and directions for understanding biological systems as a whole.

---

[3]A comprehensive list of multi-omics methods and software is available at https://github.com/mikelove/awesome-multi-omics.

2

The cycle of analysis

The *Problem, Plan, Data, Analysis, Conclusion* (PPDAC) cycle is a useful framework for answering an experimental question effectively (Figure 2.1). The mixOmics project emphasises crafting a well-defined biological question (Chapter 4), as this guides data acquisition and preparation (Chapter 8), as well as choosing appropriate multivariate techniques for analysis (Chapter 4). Although this book is focused on analysis and interpretation, careful consideration of each step will maximise a successful analytical outcome.

FIGURE 2.1: PPDAC. The *Problem, Plan, Data, Analysis, Conclusion* cycle proposed by MacKay and Oldford (2000) will guide our multivariate analysis process.

2.1 The *Problem* guides the analysis

Multivariate analysis is appropriate for large data sets where the biological question encompasses a broad domain, rather than parsing the action of a single or small number of variables. Thus, we often require a hypothesis-free investigation based on a *data-driven* approach. However, this does not imply that multivariate analysis is a fishing expedition with no underlying biological question. The experimental design, driven by a well-formulated biological question and the choice of statistical method, will ensure a successful analysis (Shmueli, 2010). Chapter 4 lists several types of biological questions that can be answered with multivariate and integrative methods.

DOI: 10.1201/9781003026860-2

2.2   *Plan* in advance

To consult the statistician after an experiment is finished is often merely to ask him to conduct a post mortem examination. He can perhaps say what the experiment died of.

– Sir Ronald Fisher, Statistician, Presidential Address to the First Indian Statistical Congress, 1938

Thoughtful experimental design is essential to ensure reproducible and replicable results as much as possible (Nichols et al., 2021). Once the biological question is formulated, care must be taken in choosing appropriate omics technologies, statistical methods, and a sufficient sample size to parse biological variation from individual differences.

2.2.1   What affects statistical power?

The underlying biological question will narrow the choice of appropriate omics technology (see Chapter 4), keeping in mind that each technology has its own artefacts and generates noise that may mask the effect of interest. The type of organism under study will also affect the amount of biological variation we expect to uncover. Similarly, the nature of the effect of interest, whether subtle, or strong, will also impact the amount of variation we can extract from the data. Taken together, these issues will affect the sample size needed to detect the effect of interest (as discussed in Saccenti and Timmerman (2016) and Lenth (2001)).

2.2.2   Sample size

Statistical power analyses are mostly valid for inferential and univariate tests but difficult to estimate when the method considered for statistical analysis does not assume any specific data distribution. As such, methods in `mixOmics` that are exploratory in nature do not fit into a classical power analysis framework. This limitation is a double-edged sword. On the one hand, any exploration on any sample size can be conducted to understand the amount of variation present in the data, and whether this variation coincides with the biological variation of interest. Such an approach can be useful for a pilot study to choose an omics technology, for example. On the other hand, a small sample size will limit any follow-up analysis: as the number of variables becomes very large, a small sample size is likely to lead to overfitting if the analysis goes beyond exploration (discussed further in Section 1.4).

Appropriate sample sizes can be calculated if pilot data are available. However, if the aim

TABLE 2.1: Example of a confounding factor in a poorly designed experiment. The number of samples with respect to the variable of interest (treatment A or B) and the covariate (sex) highlights a confounding factor (zeros outside the diagonal).

	A	B
F	10	0
M	0	10

TABLE 2.2: Example where the number of samples with respect to the variable of interest (treatment) and the covariate (sex) are balanced across treatments.

	A	B
F	5	5
M	5	5

is to use multivariate methods, the calculation will rely on an empirical approach using permutation tests[1].

2.2.3 Identify covariates and confounders

Covariates are observed variables that affect the outcome of interest but may not be of primary interest. For example, consider an experiment examining the effect of weight (primary variable of interest) on gene expression. Subject sex, age, ethnicity, or medication intake are all examples of covariates, that if assessed as having an effect on the data, must be taken into account in the analysis.

Confounding occurs when a covariate is intimately linked with the primary variable of interest. A typical example is when all females receive treatment A and all males treatment B (Table 2.1). In this case, we are not able to differentiate whether gene expression is affected by treatment, or sex, or both. Confounders can be avoided during the experimental planning stage by ensuring that both treatments are evenly assigned across sex (Table 2.2).

2.2.4 Identify batch effects

Batch effects refer to variation introduced by factors that are unrelated to the biological variable of interest. These factors can be technical (e.g. day of experiment, technician, sequencing run), computational (bioinformatics methods) or biological (birth dam, animal facility), as described in Figure 2.2 and Table 2.3. The nature of the variation is defined as *systematic* across batches, however, depending on the omics technology, this may not be the case. For example, specific genes or micro-organisms might be more affected by one type of batch compared to others (Wang and Lê Cao, 2019).

Batch effects can be mitigated with an appropriate experimental design so that their variation does not overwhelm the biological variation. When strong batch effects cannot be avoided, methods exist to correct or to account for batch effects (see Section 8.1).

---

[1]Permutation tests build a sampling distribution based on the existing data by resampling them randomly.

TABLE 2.3: Example of a batch factor. Number of samples with respect to the variable of interest (treatment) and the batch (sequencing run). Such a table gives a better understanding of the experimental design when the batch information is recorded.

	A	B
Run1	5	1
Run2	5	9

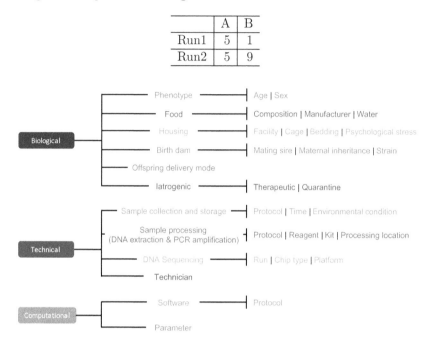

FIGURE 2.2: Potential sources of batch effects. Examples of factors that may affect the quality of microbiome data, from Wang and Lê Cao (2019).

2.3 *Data* cleaning and pre-processing

Prior to analysis, the data sets should be normalised and pre-filtered using appropriate techniques specific to the type of omics technology platform used. We briefly describe these steps and further discuss relevant techniques in Chapter 8.

2.3.1 Normalisation

High-throughput biological data often results in samples with different sequencing depths (RNA-sequencing) or overall expression levels (microarrays) due to technological platform artefacts rather than true biological differences. Thus, data must be normalised for samples to be comparable to one another. Moreover, highly-expressed genes may take up a large part of sequenced reads in an experiment, with fewer reads remaining for other genes, leading to a possibly incorrect conclusion that the latter genes have log expression levels (Robinson and Oshlack, 2010). Many normalisation techniques have been proposed that are platform-specific and constantly evolve with the latest technological, computational, and statistical advances.

2.3.2 Filtering

Whilst most multivariate methods can manage a large number of molecules (several tens of thousands), it is often beneficial to remove those that are not expressed, or vary very little, across samples. These variables are usually not relevant for the biological problem, add noise to the multivariate analyses, and may increase computational time during parameter tuning.

2.3.3 Missing values

In biological terms, missing values refer to data that are not measured (see Section 1.4). However, in practice, the definition of *missing* is often unclear. For example, a value might be missing because it is not relevant biologically or because it did not pass the detection threshold in a mass spectrometry experiment. Therefore, the data analyst must make the choice of setting the missing value to a numerical 0 (i.e. the 'absence' is biologically relevant), or 'NA' (i.e. technically missing), keeping in mind that these decisions will affect the statistical analysis.

2.4 *Analysis*: Choose the right approach

Data analysis can be descriptive, exploratory, inferential, or include modelling and prediction. A well-framed biological question will clarify which type of analysis is better suited to address a biological problem.

2.4.1 Descriptive statistics

Descriptive statistics precede any other type of analysis. They help to anticipate future analytical challenges and obtain a basic understanding of the data. Descriptive statistics solely describe the properties or characteristics of the data without making any assumptions that the data were sampled from a population. In descriptive analysis, we use summary statistics or visualisations to either obtain an initial description of the data, or investigate a particular aspect of the data. Univariate summary statistics can include calculating the sample mean and standard deviation, or graphing boxplots of a single variable. Bivariate summary statistics can include the calculation of a correlation coefficient, or a scatterplot representing the expression values of two genes. When dealing with a large number of variables, such summaries can become cumbersome and difficult to interpret, which may be overcome by using exploratory methods.

2.4.2 Exploratory statistics

Exploratory statistics aim to summarise and provide insights into the data before proceeding with the analysis. This step does not lay any emphasis on an underlying biological hypothesis and may not require a specific data distribution. Typically, such analyses can be conducted to better understand the major sources of variation in the data by using dimension reduction methods such as Principal Component Analysis (Jolliffe, 2005). More sophisticated integrative methods can also be applied to highlight the correlation structure between variables from different data sets, for example with Canonical Correlation Analysis (Vinod, 1976), or Projection to Latent Structures, also called Partial Least Squares (Wold, 1966).

2.4.3 Inferential statistics

Inferential statistics allow us to infer from a sample taken from a population. They go together with univariate statistics. If we assume that a sample is representative of the population it was drawn from, then the conclusions reached from the statistical test should generalise to the whole population. In univariate methods, we conduct hypothesis testing, calculate test statistics and compare them to a theoretical statistical distribution in order to infer that our conclusions indeed could be true for the whole population.

In the omics era, statistical inference is difficult to achieve and is often unreliable due to an insufficient sample size. Inferential statistics can be achieved with multivariate methods, but only when the number of samples, or individuals, is much larger than the number of variables, and when the multivariate methods fit into an inferential statistical framework.

2.4.4 Univariate or multivariate modelling?

Another approach in data analysis is to fit a statistical model, i.e. a mathematical and often simplified formula, to the data. The ultimate aim of modelling is to make predictions based on the then-fitted model to a new set of data, when applicable. Here is an example of an equation for a simple linear regression that models a dependent variable (or outcome variable, e.g. a phenotype) with an explanatory variable (also referred to as an independent variable, or predictor, e.g. the expression levels of a gene):

$$\text{phenotype} = \text{intercept} + a * \text{gene expression} \tag{2.1}$$

In model (2.1), a is a coefficient representing the slope of the line fitted if we were to plot the gene expression levels along the x-axis and the phenotype along the y-axis (Figure 2.3). Therefore, we assume the relationship between the dependent variable phenotype and the independent variable gene expression is linear. Using an *ordinary least squares* approach (OLS[2]), we can estimate the intercept or offset value (here estimated to -58.95), and the slope a of this linear relationship (here estimated to 6.71). We often refer to this model as a *univariate linear regression*.

A *multivariate linear regression* model includes several explanatory variables or predictors. For example in a transcriptomics experiment, if we had measured the expression levels of P genes:

$$\text{phenotype} = \text{intercept} + a_1 * \text{gene}_1 + a_2 * \text{gene}_2 + \cdots + a_P * \text{gene}_P \tag{2.2}$$

where the intercept and regression coefficients $(a_1, a_2, \ldots, a_P)$ can be estimated using the

---

[2]OLS estimates the unknown parameters in a linear regression model (e.g. the *slope* (a) and the *intercept*) by estimating values for these parameters that minimise the sum of the squares of the differences between the observed values of the variable and those predicted by the linear equation.

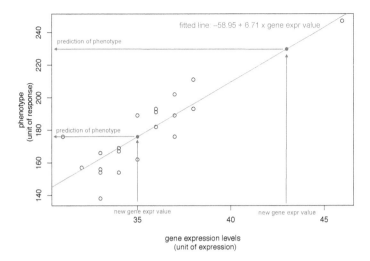

FIGURE 2.3: Example of a linear regression fit between the expression levels of a given gene (along the x-axis) and the phenotype value (along the y-axis) for 20 individuals. The model from the linear regression equation is fitted to the data points and is represented as a blue line using the Ordinary Least Squares method. Red full circles represent new gene expression values from which we can predict the phenotype of two new individuals, based on this fitted model.

OLS method. However, when the number of predictors P becomes much larger than the number of samples, the algebra to estimate these parameters becomes complex and sometimes numerically impossible to solve due to multi-collinearity (see Section 1.4). A scatterplot representation of all possible pairs of variables is often impractical and not easy to interpret.

The methods available in `mixOmics` use similar types of models as Equation (2.2), but instead of a separate OLS approach, we use *Partial* Least Square approaches to manage high dimensional data, as described in detail in Chapter 5. We also use other techniques such as ridge and lasso regularisation to manage the large number of variables (see Section 3.3).

2.4.5 Prediction

Predictive statistics refers to the process of applying a statistical model or data mining algorithm to *training* data to make predictions about new *test* data, often in terms of a phenotype or disease status of those new samples[3]. In the univariate regression model in Equation (2.1) that is fitted on training data, let us assume we also measured the expression levels of that same gene but in another cohort of patients. We can then predict the phenotype of those new patients based on that same equation. The new gene expression values for the new samples are represented in full red circles in Figure 2.3, with their y-axis values corresponding to the predicted phenotype.

---

[3]In our methods we focus on point estimate, i.e. a single value estimate of a parameter rather than interval estimate, i.e. a range of values where the parameter is expected to lie.

2.5 *Conclusion* and start the cycle again

A well-structured experiment will aid in forming a coherent and logical conclusion. The exploratory numerical summaries and graphs support and clarify the results of the main analyses. The conclusion is an opportunity to discuss the results in biological terms, the strengths and weaknesses of the statistical approach used, and to plan ahead further follow-up analyses and validation before the cycle starts again.

2.6 Summary

The best possible results for scientific research occur when an experiment is well planned. A useful framework for experimental design is to consider the PPDA cycle, where a defined biological question is formulated, and the experiment is planned in advance to minimise confounding variables and maximise the ability to detect a biological effect amidst noise. Appropriate filtering and normalisation should be performed according to the technological platform used. An in-depth analysis should include descriptive statistics to examine initial patterns or to identify possible outliers in the data[4], and appropriate exploratory, inferential or predictive statistics to answer the biological question. A coherent conclusion from these steps can help clarify the results and thus potentially lead to new knowledge, and new questions that will start the cycle of research again.

---

[4]Outliers should not necessarily be discarded, as they can bring to light some interesting aspects of a data set, from both substantive and technical perspectives.

3

Key multivariate concepts and dimension reduction in `mixOmics`

In `mixOmics`, we aim to shift the univariate analysis paradigm towards a unifying multivariate framework and mitigate many of the challenges arising from high-throughput omics data. The multivariate methods developed in this package use matrix factorisation techniques and regression penalties to reduce the dimensions of the data and identify multi-omics molecular signatures.

This chapter introduces important statistical measures, such as *variance*, *covariance*, and *correlation*, which are used in our methods to calculate *components* and *loading vectors*. These form the basis of `mixOmics`' comprehensive visualisation techniques.

3.1 Measures of dispersion and association

Measures of association are used to quantify a relationship between two or more variables. We first introduce the concept of *variance* as a measure of dispersion of a *random variable* before examining the difference between *covariance* and *correlation* matrices.

3.1.1 Random variables and biological variation

In Statistics, we consider the measurement of biological molecules as *random variables*: each variable (e.g. gene, protein) may be associated with a range of possible numerical values (e.g. gene expression levels, protein concentrations). Each value is associated with a probability of occurring: some values have a higher probability of occurring in some individuals, while other values will have a lower probability of occurring, depending, for example, on a particular treatment or diet. The *biological variation* is the source of variation we expect to observe from a well-planned experiment.

However, biological data may also contain variation from non-biological sources, for example, technical factors such as different sequencing runs, or sample handling by different technicians, as discussed in Section 2.2. This *unwanted variation* may outweigh the interesting biological variation from the research experiment. Therefore, examining variation in the data is the first step in the exploration and mining of high-throughput biological data.

DOI: 10.1201/9781003026860-3

3.1.2 Variance

Variance is a key concept in data analysis. For example, we may be interested in how the expression levels of a given gene vary across a patient cohort, or how a protein's concentration varies over time. Numerically, variance is a measure of the dispersion, or spread, in a variable's possible values.

We denote the random variable x (e.g. the expression levels of a given gene in a given population of size N), and μ_x the mean of that variable. The population variance is:

$$\text{Var}(x) = \frac{1}{N} \sum_{i=1}^{N} (x_i - \mu_x)^2 \tag{3.1}$$

where N is the population size, and x_i corresponds to the gene expression levels for individual i.

As we do not have access to the measurements from a whole population, we calculate instead the *sample variance* estimated from our data. This consists of calculating the difference between the variable's value and its sample mean, then squaring the result. The *standard deviation* is the square root of the variance $\sqrt{\text{Var}(x)}$.

A large variance indicates that most of the variable's values are spread out far away from their sample mean. A small sample variance implies the variable's values are clustered close to the sample mean. Depending on our biological question, we may be interested in either a large or small sample variance. For example, we may observe a large variance in the expression levels of a given gene across samples that have undertaken different treatments. Conversely, we expect to observe a small variance in gene expression if we consider a homogeneous group of samples (i.e. within a given treatment).

Note:

- *In later chapters, instead of using the notation x, we will use X^j to denote the variable in column j stored in the matrix X, where X is of size $N \times P$, $j = 1, \ldots, P$.*

3.1.3 Covariance

The covariance is a bivariate measure of association between two random continuous variables. For example, we might be interested in quantifying the joint variability between two genes measured on a population. We denote x and y as two random variables from a given population, and their respective means μ_x and μ_y. We define the population covariance as:

$$\text{Cov}(x, y) = \frac{1}{N} \sum_{i=1}^{N} (x_i - \mu_x)(y_i - \mu_y) \tag{3.2}$$

where N is the population size, and x_i (y_i) corresponds to one or the other gene expression levels for individual i.

From Equations (3.1) and (3.2), we can see that the population covariance is a generalisation of the variance for two random variables, where we multiply the differences between each variable's value and their respective means. Similar to the concept of sample variance, we estimate the sample covariance between two variables we have measured from our data, except that this time we divide by $N-1$ rather than N in Equation (3.2) to obtain an unbiased estimate.

If most variables' values are greater or less than their respective means, then the product of the differences is positive, and we can conclude on a *positive association*. If the value of one variable is greater than its mean but the value of the other variable is less than its mean, then the covariance is negative, i.e. a *negative association*. A null covariance indicates that either one or both variables' values are not different from their respective mean, suggesting *no association*.

We can group these variance and covariance measures into a sample *variance-covariance matrix*, where elements on the diagonal are the sample variances of each variable, and the elements outside the diagonal are the sample covariances. Here are two examples: $\boldsymbol{A}$ contains two variables denoted as subscripts 1 or 2, while $\boldsymbol{B}$ contains three variables denoted as subscripts $1, 2$, or 3:

$$\boldsymbol{A} = \begin{pmatrix} \mathrm{Var}_1 & \mathrm{Cov}_{1,2} \\ \mathrm{Cov}_{2,1} & \mathrm{Var}_2 \end{pmatrix} \qquad \boldsymbol{B} = \begin{pmatrix} \mathrm{Var}_1 & \mathrm{Cov}_{1,2} & \mathrm{Cov}_{1,3} \\ \mathrm{Cov}_{2,1} & \mathrm{Var}_2 & \mathrm{Cov}_{2,3} \\ \mathrm{Cov}_{3,1} & \mathrm{Cov}_{3,2} & \mathrm{Var}_3 \end{pmatrix} \qquad (3.3)$$

Note that the variance-covariance matrix is symmetrical with respect to the diagonal. For example in Equation (3.3), we have $\mathrm{Cov}_{1,2} = \mathrm{Cov}_{2,1}$.

The covariance measures the degree of association of two variables that may have different units of measurements, or different scales. Therefore, the difference in magnitude in a covariance matrix does not inform about the strength of the association. In some cases, it may be better to standardise the variables so that their mean is 0 and variance is 1, as detailed in Section 8.1 or alternatively, to use correlation values.

3.1.4 Correlation

The correlation is a standardised covariance measure. Using the same notations as in Equation (3.2), we define the population correlation as:

$$\mathrm{Corr}(\boldsymbol{x}, \boldsymbol{y}) = \frac{\mathrm{Cov}(\boldsymbol{x}, \boldsymbol{y})}{\sqrt{\mathrm{Var}(\boldsymbol{x})}\sqrt{\mathrm{Var}(\boldsymbol{y})}} \qquad (3.4)$$

where we divide the population covariance by the product of the population standard deviations of each of the two variables. The sample correlation is estimated from the sample variance and sample covariance from the data.

As the correlation is standardised, its value is bound between -1 and $+1$. A correlation close to $+1$ indicates a strong *positive association*, and close to -1 a strong *negative association*. A correlation close to 0 indicates that the two variables are *uncorrelated*. For example, two genes are positively correlated if their expression levels both increase across samples, but they are negatively correlated when the expression of one gene increases while the expression of the second gene decreases.

The Pearson product-moment correlation coefficient (or Pearson correlation) is the most popular measure of correlation to measure a linear relationship between two variables. However, it is very sensitive to outlier data points, or non-linear relationships. The other more robust and non-parametric alternative is the Spearman's rank correlation coefficient, which is the Pearson correlation coefficient calculated on the ranks of the variable values rather than the variable values themselves[1].

In practice, the association between the two variables can be understood by graphing them on a scatterplot. However, this is not practical when dealing with a large number of variables and certainly does not represent a multivariate description of the data. Moreover, as correlations work pairwise, we obtain a large number of correlations, many of which can be spurious and unrelated to the biology of interest (Calude and Longo, 2017). Finally, keep in mind the adage 'correlation does not imply causation', as in our context, we are only seeking for associations between variables.

3.1.5   Covariance and correlation in `mixOmics` context

In `mixOmics`, to study the association between different types of variables, we measure the covariance or correlation between *artificial* variables (*components*), which are linear combinations of the original variables. Unless specified, we will refer to the correlation measure as the Pearson correlation coefficient.

3.1.6   `R` examples

We illustrate the different measures of association on the small data set `linnerud` (Tenenhaus, 1998) that measures three exercise variables ($P = 3$) on twenty middle-aged men ($N = 20$). We first start by loading and preparing the data from the package:

```
library(mixOmics)
data("linnerud")
# exercise data only
data <- linnerud$exercise
```

The sample variance-covariance matrix is a square, symmetric matrix with 3 rows and 3 columns. The elements on the diagonal indicate the variance of each variable:

```
cov(data)
```

```
##          Chins Situps  Jumps
## Chins    27.94  230.1  134.4
## Situps  230.11 3914.6 2147.0
## Jumps   134.38 2147.0 2629.4
```

We observe that the variables' variance on the diagonal differ from one another as the variables are not standardised. Outside the diagonal, the values range from 134.4 to 2147.

The sample correlation matrix is the same size as the variance-covariance matrix, but with values of 1 on the diagonal (perfect correlation within a variable):

```
cor(data, method = 'pearson')
```

---

[1] In this book, we will refer to correlation as Pearson correlation.

```
##          Chins Situps  Jumps
## Chins   1.0000 0.6957 0.4958
## Situps  0.6957 1.0000 0.6692
## Jumps   0.4958 0.6692 1.0000
```

The correlation coefficients range from 0.4958 to 0.6957 if we disregard the diagonal. Comparing both outputs, we note that the maximal values do not correspond to the same pair of variables (Situp – Chins in the correlation matrix, and Situps – Jumps in the variance-covariance matrix) because of the standardisation performed in the correlation matrix.

To summarise, both *covariance* and *correlation* measure how variables change or covary together, but their scales differ:

1. The correlation is obtained by dividing the covariance by the standard deviation of both variables to remove units of measurement.

2. The covariance is an unstandardised version of their correlation. As such, the magnitude of the covariance can be difficult to interpret (as it depends on the magnitude of the variables).

3. Contrary to the correlation coefficient, the covariance coefficient does not range between -1 or 1 and is unbounded. Similar to a correlation coefficient, a value of 0 indicates no linear relationship and the sign of the covariance indicates the trend of the relationship between two variables.

4. Outlier values in variables can distort both (Pearson) correlation and covariance coefficients.

3.2 Dimension reduction

Matrix factorisation forms the basis of many dimension reduction techniques. It decomposes a data matrix into a smaller set of 'factors' or matrices. For example, consider the number 15, which can be decomposed into the numbers 3×5 or 5×3. In this example, 3 and 5 are different pieces of information that multiply together to form our original number of 15. Similarly, we can decompose a matrix into a number of smaller matrices, each containing different information, that when multiplied together, will re-form the original matrix.

3.2.1 Matrix factorisation

Figure 3.1 illustrates the decomposition of a matrix into two sub-matrices representing different kinds of information. One sub-matrix represents information about the rows (samples) of the original matrix, and the second sub-matrix represents information about the columns (variables) of the original matrix.

Different dimension reduction methods use different statistical criteria to define each of these sub-matrices so that they reflect particular types of underlying patterns in the data (see the reader-friendly review of matrix factorisation in omics data from Stein-O'Brien et al.

(2018)). In Part III of this handbook, we will detail each statistical criterion used to define the matrix factorisation technique specific to each method in `mixOmics`.

3.2.2 Factorisation with components and loading vectors

Matrix factorisation techniques are especially useful when managing large, noisy, and multicollinear data, as the vectors from the sub-matrices aim to summarise most of the information from the data.

In the example depicted in Figure 3.1, we have decomposed the original data matrix into a sub-matrix of two vertical vectors (or *components*) that summarise the data at the sample level, and a sub-matrix of two horizontal vectors (or *loading vectors*) that summarise the data at the variable level. The components can be regarded as artificial variables, and are constructed so that they aggregate or combine the information of all P variables into two dimensions. Thus, whilst the original data matrix is described by a P-dimensional space (i.e. each sample is described by P parameters, for example, P genes), we can now describe the samples using a much smaller dimensional space, for example, a two-dimensional space if we choose two components to summarise the data.

The components aggregate the variables by using the information from the loading vectors, which contain the weight information assigned to each variable. In this example, each component is associated with a loading vector. By multiplying the components and loading vectors, we can reconstruct an approximation of the original matrix. The more components (and corresponding loading vectors) we extract, the closer we might get to the original matrix, however, this would result in a less effective dimension reduction. Chapter 5 will give additional details on how loading vectors and components are calculated.

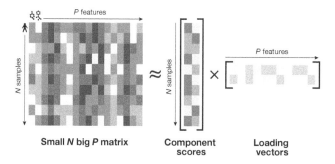

FIGURE 3.1: Example of matrix factorisation of a data matrix with N samples in rows and P variables in columns, into a set of two components and two corresponding loading vectors. These sets of vectors are obtained using algebra calculations further detailed in Chapter 5. The dimensions of P variables are reduced to two-dimensions, artificial variables, or components.

3.2.3 Data visualisation using components

The components obtained from matrix factorisation techniques provide an insightful data representation at the sample level (Figure 3.2). Each sample's coordinate on the x and y axes are the elements or score values from the first and second components. The scatterplot

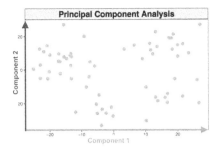

FIGURE 3.2: Sample plot. Visualisation of the samples (dots) into a low-dimensional space. The samples' coordinates on the x and y axes correspond to their values from the first two components. This type of plot represents a *projection* of the samples (originally described by P variables, or P-dimensional space) into a 2D-space spanned by the two components.

obtained is a *projection of the samples* from a P-dimensional space into a smaller subspace of two dimensions – the space spanned by the components.

These sample plots are the most common type of plot from matrix factorisation methods and allow us to observe clusters of samples that may share similar biological characteristics (based on their component values), identify potential sample outliers, and assess for the unexpected presence of confounding variables, or laboratory and platform effects. Chapter 6 describes other types of plots available in `mixOmics`.

3.3   Variable selection

The `mixOmics` package stands out from other R packages for omics data integration due to its focus on variable selection, also called *feature selection*. Examining the components after dimension reduction only provides the first level of understanding of the data at the sample level. We can also extract additional information concerning which variables are key to explain the variance or covariance of the data.

Loading vectors reflect the importance of each variable in defining each component. In `mixOmics`, we have developed novel methods that apply different types of regularisation techniques, or penalty terms, on the loading vectors. These methods enable us to either manage numerical limitations resulting from large data sets during the matrix factorisation step (*ridge* regularisation), or identify variables that are influential in defining the components (*lasso* regularisation). The latter can be considered as an extra dimension reduction step, as irrelevant variables are ignored when calculating the components. These techniques necessitate choosing regularisation, or *penalty* parameters using *tuning* functions, as detailed in Chapter 7 and subsequent chapters in Part III. Here we give a broad introduction to the family of penalties implemented in the methods.

3.3.1   Ridge penalty

The ℓ_2-norm or ridge penalty consists of shrinking the coefficients of the non-influential variables in the loading vectors towards small but non-zero values (Hoerl, 1964). We use this underlying principle in regularised Canonical Correlation Analysis (González et al., 2008) and regularised Generalised CCA (Tenenhaus and Tenenhaus, 2011) to numerically manage large data sets and their covariance matrices.

3.3.2   Lasso penalty

The ℓ_1-norm or lasso penalty (Least Absolute Shrinkage and Selection Operator) is a popular type of penalty for large data sets (Tibshirani, 1996). In `mixOmics`, we loosely refer to methods using lasso penalisation as *sparse methods*, as lasso shrinks many of the variables' coefficients in the loading vectors to exactly zero. These variables are deemed biologically irrelevant, whilst the remaining variables with non-zero coefficients represent our identified molecular signature. Lasso regularisation is used in almost every method in the package for variable selection, with the exception of regularised Canonical Correlation Analysis (rCCA) and regularised Generalised Canonical Correlation Analysis (rGCCA).

3.3.3   Elastic net

Elastic net is a combination of ℓ_1 and ℓ_2 penalties to address the limitations of both approaches. When the number of variables selected is very small, lasso penalty tends to select uncorrelated variables, while ridge penalty manages correlated variables but does not perform variable selection. Elastic net can select groups of correlated variables (Zou and Hastie, 2005), and is implemented in sparse Generalised Canonical Correlation Analysis (sGCCA) (Tenenhaus et al., 2014) in `mixOmics`, as we detail in Chapter 13.

3.3.4   Visualisation of the selected variables

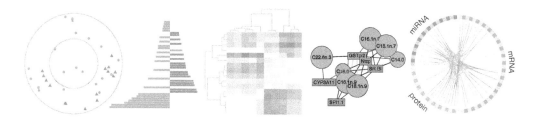

FIGURE 3.3: Illustration of different variable plots in `mixOmics` to represent the variables' correlation structure, their importance, and the relationship between samples and variables. When using sparse methods, only the selected variables are displayed. These graphics are explained in more detail in Chapter 6 and relevant Chapters in Part III.

We propose different types of plots in `mixOmics` to interpret the role and importance of the variables, their correlation structure, as well as the relationship between samples and variables (Figure 3.3). These plots are based on either the loading vectors, the components, or both types of information to estimate associations between variables. When sparse methods are used, only the selected variables are displayed. We give more details about the variety of plots available in Chapter 6.

3.4   Summary

The `mixOmics` package proposes a unifying multivariate framework to reduce the dimensions of high-throughput biological data, both at the sample level, by defining components as artificial variables, and at the variable level, by using penalties to identify key variables influencing the biology of interest. Subsequent chapters will detail how the methods maximise specific criteria such as the variance of components, or the covariance or correlation between components, to answer specific biological questions.

Dimension reduction is achieved through matrix factorisation techniques, which decompose matrices into sets of components and loading vectors. Components are artificial variables that capture underlying structure in the data, whereas loading vectors indicate the importance of the variables that are used to calculate the components. The calculation of the components and loading vectors will be elaborated in Chapter 5. Variable selection is achieved by applying lasso regularisation to the loading vectors so that components are only defined by the important variables that explain the structure of the data.

Graphical plots that represent the projection of the samples into the space spanned by the components, and the information from the loading vectors, can then be used to give more insight into the data projected into a smaller dimension. These plots will be described in more detail in Chapter 6.

4

Choose the right method for the right question in *mixOmics*

The `mixOmics` package offers a wide range of multivariate methodologies that employ dimension reduction and feature selection to address different types of biological questions and guide further analytical investigations. Before we delve into the specific methods and techniques in the package, it is important to outline the different kinds of analyses, and the specific biological questions our methods can answer, based on how the data are structured. We illustrate these questions on the exemplar data sets available in `mixOmics`.

4.1 Types of analyses and methods

Currently, nineteen multivariate projection-based methods are implemented in `mixOmics`. Twelve methods have similar names, and call an extension of the PLS algorithm from Tenenhaus et al. (2014). They have been named to represent different combinations of analysis frameworks. For example, `mint.block.splsda` extends `'pls'` to perform a sparse (`'s'`) discriminant analysis (`'da'`) on multiple data sets (`'block'`) measured on the same p variables (`'mint'`). The remaining five statistical methods are PCA and sparse PCA, Independent PCA (Yao et al., 2012), regularised CCA and its generalisation to multiple data sets rGCCA (Tenenhaus and Tenenhaus, 2011, Tenenhaus et al., 2014). Each statistical method returns a list of outputs which are essential for visualisation.

Method choice depends on the biological problem, the number of data sets, and whether the dependent variable is quantitative, qualitative, or is presented as a matrix of more than one dependent variable. Figure 4.1 provides an overview of the most widely used methods in `mixOmics`. We then introduce the different types of analyses and corresponding methods available in the package.

4.1.1 Single or multiple omics analysis?

The best approach for constructing a methodical integrative data analysis is to first analyse each data set separately, and decide whether an unsupervised or supervised analysis is required. Single omic analysis often constitutes the preliminary step to understanding the data prior to the integration of multiple data sets.

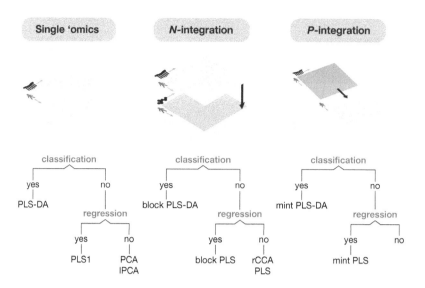

FIGURE 4.1: Different types of analyses with `mixOmics`. The biological questions, the number of data sets to integrate, and the type of response variable, whether qualitative (classification), quantitative (regression), one (PLS1), or several (PLS) responses, all drive the choice of analytical method. All methods featured in this diagram include variable selection except `rCCA`. In $N-$integration, `rCCA` and `PLS` enable the integration of two quantitative data sets, whilst the `block PLS` methods (that derive from the methods from Tenenhaus and Tenenhaus (2011)) can integrate more than two data sets. In $P-$integration, our method MINT is based on multi-group PLS (Eslami et al., 2014).

4.1.2 $N-$ or $P-$integration?

Our integrative frameworks range from two omics up to multiple omics. Within the multiple omics analyses we consider $N-$ and $P-$integration: $N-$integration combines different biological molecules, or variables, measured on the same N individuals or samples to characterise a biological system with a holistic view. $P-$integration combines the same P molecules (e.g. genes with the same identifier) measured across different data sets generated from several laboratories but interested in the same biological question to increase sample size or to identify a common signature between independent studies. Note that multi-study data integration ($P-$integration in `mixOmics`) has been coined 'horizontal integration' (Tseng et al., 2012), while multi-omics data integration ($N-$integration in `mixOmics`) is 'vertical integration'. However, since our data are transposed with samples in rows and variables in columns in our methods, these terms do not intuitively reflect our frameworks (as illustrated in Figures 13.1 for $N-$ and 14.1 for $P-$ integration).

For both types of integrative framework, the overarching aim is to describe patterns of agreement or correlation between these different data sets, or to explain a biological outcome or *phenotype*[1]. Examples of phenotypic variables are clinical measures, or lipid concentration, depending on the biological question.

---

[1]A phenotype is often defined as the observable characteristics of an individual that are influenced by genotype and environmental interactions.

4.1.3 Unsupervised or supervised analyses?

4.1.3.1 Unsupervised analyses

Unsupervised analyses can identify patterns in data, such as clusters of samples or variables, *without taking into account a priori* information about the sample groups, or outcome information (e.g. disease status, cancer subtype, treatment intake). These methods can visualise the data in an unbiased manner, and inform how samples 'naturally' form subgroups based on how similar their expression profiles are according to a statistical criterion of interest (e.g. variance). Unsupervised analysis often refers to *data exploration* and *data mining*.

In `mixOmics` we provide unsupervised methods for single, and the integration of multiple, data sets. Sparse methods also allow for the identification of subsets of relevant (and possibly correlated) variables. Unsupervised methods for exploratory analyses include:

- Principal Component Analysis (Jolliffe, 2005) `pca` and `spca`,
- Independent Principal Component Analysis (Yao et al., 2012) `ipca` and `sipca`,
- regularised Canonical Correlation Analysis (González et al., 2008) `rcca`.

4.1.3.2 Supervised analyses

Supervised analyses are the counterpart of unsupervised analyses and often refer to broad types of techniques used in the Machine Learning field. They are used for biomarker discovery and predicting an outcome. In *classification* analysis, there is one outcome variable that is categorical and expressed as class membership of all samples. In *regression* analysis, the response variable (s) is (are) continuous. Both analyses aim to model the relationship between the variables measured and the outcome.

Methods for regression analysis include (Figure 8.5):

- Partial Least Squares regression (PLS, also known as Projection to Latent Structures Wold (1966)) `pls`, `spls`, `block.pls` and `block.spls`,
- multi-group PLS (MINT) (Eslami et al., 2013) `mint.block.pls` and `mint.block.spls`.

Methods for classification analysis in `mixOmics` are popular as we can use cross-validation, or an external test set, to assess how well the model generalises to (artificially created) new data, as we describe in Chapter 7. Methods for single and multiple data set analyses include:

- PLS-Discriminant Analysis (PLS-DA) (Boulesteix, 2004; Nguyen and Rocke, 2002a,b) `plsda` and `splsda`,
- DIABLO (Singh et al., 2019) `block.plsda` and `block.splsda`, as an extension from Tenenhaus et al. (2014),
- Multi-group PLS-DA (Rohart et al., 2017a) `mint.block.plsda` and `mint.block.splsda`.

Note:

Other wrapper methods have been implemented for regularised Generalised Canonical Correlation Analysis (rGCCA) wrapper.rgcca from Tenenhaus and Tenenhaus (2011) and sparse Generalised CCA wrapper.sgcca from Tenenhaus et al. (2014) in mixOmics. Both methods are based on a PLS algorithm (Tenenhaus et al., 2015). They are briefly presented in Section 13.A and can fit into an unsupervised or a supervised (classification or regression) framework.

4.1.4   Repeated measures analyses

In a repeated measures experiment (e.g. cross-over design), different treatments or conditions are applied on the same subjects. In this type of design, the individual variation can be much greater than the subtle variation we try to uncover from the treatment. A simple PCA can highlight whether this is the case (see Figure 4.2 **(a)** in Appendix 4.A.1). A multilevel approach can help to highlight the subtle differences between conditions *within subjects* from the biological variation between subjects (see Figure 4.2 **(b)** in Appendix 4.A.1).

We consider this approach as an additional pre-processing step to the data, as it consists of extracting the within individual variation that is of interest from the total variation in the data (Liquet et al., 2012)[2].

The multilevel approach may also apply for a time course experiment with a very small number of time points or repeats (two to three, e.g. a before and after treatment), see for example, Lee et al. (2019). Methodological details about the multilevel decomposition can be found in Appendix 4.A.1 with further illustrations using `mixOmics`.

4.1.5   Compositional data

Data of compositional nature arise from high-throughput sequencing technologies where the number of reads that can be sequenced is limited by the capacity of the instrument (Gloor et al., 2017). This is the case for RNA-seq, 16S rRNA gene data, shotgun metagenomics, and methylation data, for example. In addition, read counts are often converted into relative abundance or normalised counts. Thus, either a finite amount of sequences, or the transformation of proportions, results in variables with a constant sum. Compositional data reside in a bounded space that is not Euclidean (Aitchison, 1982). As most statistical methods assume that data live in an unbounded (Euclidean) space, the analysis may lead to spurious results as the independence assumption between predictor variables is not met (Lovell et al., 2015). Microbiome data has been described as compositional also due to ecological reasons, because of the possible co-existence/dependence of specific types of microorganisms.

A growing list of references advocates against the use of conventional statistical methods for the analysis of compositional data. One practical solution is to transform compositional data into Euclidean space using centered log ratio transformation (CLR) before applying standard univariate or multivariate methods (Aitchison, 1982; Fernandes et al., 2014; Lovell et al., 2015). The CLR transformation of count data in `mixOmics` is described in Appendix 4.B.

---

[2]An internal argument `multilevel` is available in the methods PCA and PLS-DA, as shown in Section 12.6.2. Alternatively, one can use the `withinVariation()` function to extract the within variation matrix.

4.2 Types of data

4.2.1 Classical omics

Different types of biological data can be explored and integrated with `mixOmics`. Prior to the analysis, we assume the data sets have been filtered and normalised (Chapter 8) using techniques specific for the type of omics technology platform. The methods can manage molecular features measured on a continuous scale (e.g. microarray, mass spectrometry-based proteomics, lipidomics, and metabolomics) or sequenced-based count data (RNA-seq, 16S rRNA gene, shotgun metagenomics, and methylation) that become 'continuous' after pre-processing and normalisation, as well as clinical data.

4.2.2 Microbiome data: A special case

As mentioned in the previous section, microbiome data generated with 16S rRNA gene amplicon sequencing or whole-genome sequencing are inherently compositional and require a log-ratio transformation to be appropriately analysed. Microbiome data also include a large number of zeros. As such, variables with a predominant number of zeros should be filtered out (see Section 8.1). We provide further examples at http://www.mixOmics.org.

4.2.3 Genotype data: A special case

Genotype data, such as bi-allelic Single Nucleotide Polymorphism (SNP) coded as counts of the minor allele can also fit in `mixOmics`, by implicitly considering an additive model. However, our methods are not yet able to manage SNPs as categorical variables as this requires additional methodological developments. Another option to deal with SNP genotypes is to calculate polygenic risk scores established from using data from different consortiums (e.g. from GTex, Lonsdale et al. (2013)), then consider these scores as a phenotype variable in the multivariate analysis.

4.2.4 Clinical variables that are categorical: A special case

All multivariate methods in `mixOmics` are based on a PLS-type algorithm that assumes all explanatory variables are continuous. The technique for handling categorical variables is to transform them into binary variables corresponding to each category, and then code them as 0 or 1 for each category. All binary variables are then concatenated into a dummy matrix for analysis. Note, however, that the analysis might be suboptimal, or difficult to interpret. We give an example in Appendix 4.C using the `unmap()` function provided in `mixOmics`.

4.3 Types of biological questions

Below we give examples of broad biological questions, a description of the methods in `mixOmics` that can address these questions, and detailed examples based on the data sets available in the package.

4.3.1 A PCA type of question (one data set, unsupervised)

Biological question: What are the major trends or patterns in my data? Do the samples cluster according to the biological conditions of interest? Which variables contribute the most to explaining the variance in the data?

Multivariate method: **Principal Component Analysis** (PCA) can address these questions by achieving dimension reduction and visualisation of the data (Jolliffe, 2005). The principal components are defined to uncover the major sources of variation in the data. Similarities between samples and the correlation between variables can be visualised using sample plots and correlation circle plots. Variants such as **sparse PCA** further allow for the identification of key variables that contribute to defining the components while **Independent Principal Component Analysis** (IPCA) uses ICA as a denoising process to maximise statistical independence between components. Contrary to PCA, ICA is a non-linear approach.

Example in `mixOmics`*:* The `multidrug` study contains the expression of 48 known human ABC transporters from 60 diverse cancer cell lines (Scherf et al., 2000; Weinstein et al., 1992) used by the National Cancer Institute to screen for anticancer activity. With PCA, we can identify which samples are more similar to each other than others, and whether they belong to the same type of cell line. We can further identify specific ABC transporters that may drive these similarities or differences (Chapter 9).

4.3.2 A PLS type of question (two data sets, regression or unsupervised)

Biological question: My two omics data sets are measured on the same samples, where each data set (denoted X and Y) contains variables of the same type. Does the information from both data sets agree and reflect any biological condition of interest? If I consider Y as phenotype data, can I predict Y given the predictor variables X? What are the subsets of variables that are highly correlated and explain the major sources of variation across the data sets?

Multivariate method: **Projection to Latent Structures** (a.k.a Partial Least Squares regression, PLS) and its variant **sparse PLS** can address these questions (Lê Cao et al., 2008; Wold, 1966). PLS maximises the covariance between data sets via components, which reduce the dimensions of the data. In sparse PLS (sPLS), lasso penalisation is applied on the loading vectors to identify the key variables that covary (see Section 3.3). Two variants were developed for a regression framework; the first is applied when one data set can be explained by the other (**sPLS regression mode**, Lê Cao et al. (2008)), while the second is a canonical correlation framework similar to CCA where both data sets are considered symmetrically (**sPLS canonical mode**, Lê Cao et al. (2009)).

Examples in `mixOmics`*:* The `liver.toxicity` study contains the expression levels of 3116 genes and 10 clinical chemistry measurements containing markers for liver injury in 64 rats. The rats were exposed to non-toxic, moderately toxic, or severely toxic doses of acetaminophen in a controlled experiment (Bushel et al., 2007). PLS can highlight similar information between both data sets (acetaminophen dose and time of necropsy). Sparse PLS can highlight a subset of clinical markers related to liver injury that can also be explained by a subset of gene expression levels (Chapter 10).

The `multidrug` study is revisited with the integration of 48 ABC transporters and the activity of 1,429 drugs measured from the 60 cancer cell lines (Szakács et al., 2004). PLS canonical mode highlights any agreement between transporters and drug activity variables that may explain the variation or difference between the different types of cancer cell lines. With sPLS canonical mode, we can identify subsets of transporters and drugs that are correlated with each other and are related to specific types of cell lines (Chapter 10).

4.3.3   A CCA type of question (two data sets, unsupervised)

Biological question: I have two omics data sets measured on the same samples. Does the information from both data sets agree and reflect any biological condition of interest? What is the overall correlation between them?

Multivariate method: **Regularised Canonical Correlation Analysis** (rCCA) is a variant of CCA that can manage large data sets (González et al., 2008; Leurgans et al., 1993). rCCA achieves dimension reduction in each data set whilst maximising similar information between two data sets measured on the same samples (N−integration framework). The *canonical correlations* inform us of the agreement between the two data sets that are projected into a smaller space spanned by the *canonical factors* while the sample plots obtained via the canonical factors allows to visualise of the samples in both data sets. Contrary to sPLS, rCCA does not select variables that contribute to the correlation between the data sets.

Example in `mixOmics`: The `nutrimouse` study is a murine nutrigenomics study in which the effect of five diet regimens is investigated by measuring fatty acid compositions in liver lipids and hepatic gene expression (Martin et al., 2007). Forty mice from two different genotypes undertook one of the five diets (eight mice per diet). We analyse two data sets: the expression levels of 120 genes potentially involved in nutritional problems, and the concentrations of 21 hepatic fatty acids. With CCA, we aim to extract common information between genes and fatty acids, and investigate whether this integrated information also relates to mouse genotype and diet. In addition, through correlation circle plots, we can highlight particular genes and lipids that may be associated to a particular genotype and/or diet. As introduced in Section 3.3, rCCA employs ridge penalisation, where all variables are examined in the analysis. As such, the focus is on highlighting correlations rather than selecting specific variables (Chapter 11).

4.3.4   A PLS-DA type of question (one data set, classification)

Biological question: Can I discriminate samples based on their outcome category? Which variables discriminate the different outcomes? Can they constitute a molecular signature that predicts the class of external samples?

Multivariate method: **PLS-Discriminant Analysis** (PLS-DA) is a special case of PLS where the phenotype data set is a single categorical outcome variable (Lê Cao et al., 2011). PLS-DA discriminates sample groups during the dimension reduction process and fits into a predictive framework (see Section 2.4). The variant **sparse PLS-DA** includes lasso penalisation on the loading vectors to identify a subset of key variables, or a molecular signature (Lê Cao et al., 2011). Thus, based on either all variables (PLS-DA) or a subset (sparse PLS-DA), we can predict the outcome of new samples, or samples from an artificially-created test data set using cross-validation.

Example in `mixOmics`: The Small Round Blue Cell Tumour (`srbct`) data set includes the expression measure of 2,308 genes measured on 63 samples divided into four different subtypes of cancer; Burkitt Lymphoma, Ewing Sarcoma, Neuroblastoma, and Rhabdomyosarcoma. A PLS-DA shows that the tumour subtypes can be discriminated based on the expression levels of all genes. Sparse PLS-DA can identify a subset of genes that can discriminate and accurately predict the different tumour subtypes (Chapter 12).

4.3.5 A multiblock PLS type of question (more than two data sets, supervised or unsupervised)

Biological question: The biological question is similar to the PLS question above, except that there are more than two data sets, where one data may contain phenotype variables of the same type, and other omics data sets are available on matching samples.

Multivariate method: The **multiblock PLS** variants are based on Generalised CCA and can answer these questions to various extents. The method **block.spls** is a variant of sPLS regression mode where the phenotype matrix is regressed onto the other data sets. The covariance between pairs of components corresponding to each data set is maximised. This method called **wrapper.sgcca** has been further improved to fit into `mixOmics`, and is based on **sparse GCCA** (from the `RGCCA` package, Tenenhaus et al. (2014)). **Regularised GCCA** is a variant of rCCA for more than two data sets (Tenenhaus and Tenenhaus, 2011). Similarly to CCA, the intent is to maximise the correlation between pairs of components corresponding to each data set to uncover common information. The ridge regularisation enables us to manage a large number of variables and therefore does not include variable selection (see Section 3.3). All data sets are considered symmetrically in rGCCA. The method is available as **wrapper.rgcca** in `mixOmics` and originates from the `RGCCA` package.

Some of these methods are still being developed, refer to our website at http://www.mixOmics.org for details and application cases. The basis of GCCA is detailed in Chapter 13.

4.3.6 An $N-$integration type of question (several data sets, supervised)

Biological question: Can I discriminate samples across several data sets based on their outcome category? Which variables across the different omics data sets discriminate the different outcomes? Can they constitute a multi-omics signature that predicts the class of external samples?

Multivariate method: **multiblock PLS-DA** is a variant of sGCCA that integrates several data sets measured on the same samples in a supervised framework (Singh et al., 2019). Dimension reduction is achieved so that the covariance between pairs of data sets is maximised through PLS components. Lasso penalisation applied to each omics loading vector allows for variable selection and the identification of a correlated and discriminant multi-omics signature. We refer to this method as `DIABLO`.

Example in `mixOmics`: The `breast.TCGA` study includes a subset of the data available from The Cancer Genome Atlas (Cancer Genome Atlas Network et al., 2012). It contains the expression or abundance of three omics data sets: mRNA, miRNA, and proteomics, for the same 150 breast cancer samples subtyped as Basal, Her2, and Luminal A as the training set. The test set used for prediction assessment includes 70 samples, but only for mRNA and miRNA. With multi-block sPLS-DA we can identify a highly correlated multi-omics signature that can also discriminate and predict the different tumour subtypes on the test samples (Chapter 13).

4.3.7 A $P-$integration type of question (several studies of the same omic type, supervised or unsupervised)

Biological question: I want to analyse different studies related to the same biological question using similar omics technologies. Can I combine the data sets while accounting for the variation between studies? Can I discriminate the samples based on their outcome category? Which variables are discriminative across all studies? Can they constitute a signature that predicts the class of external samples?

Multivariate method: **group-PLS** enables us to structure samples into groups (here studies) in a PLS model (Eslami et al., 2014). The method MINT is a PLS-DA variant that integrates independent studies measured on the same P variables (e.g. genes) to discriminate sample groups (Rohart et al., 2017b). Group-PLS seeks for common information across data sets where systematic variance may occur within each study. Lasso penalisations are used to identify key variables across all studies that discriminate the outcome of interest.

Example in `mixOmics`: The `stemcells` data set contains the expression of a subset of 400 genes in 125 samples from four independent transcriptomic studies. Samples of three stem cell types are considered: human fibroblasts, embryonic stem cells and induced pluripotent stem cells (Rohart et al., 2017a). MINT enables the identification of a robust gene signature across all independent studies to classify and accurately predict the stem cell subtypes (Chapter 14).

4.4 Examplar data sets in `mixOmics`

As we highlight in this chapter, most omics studies focus on a particular type of question that can be addressed with a *specific type of method*. We summarise most of the example data sets available in `mixOmics` and the methods appropriate for their analysis in Table 4.1, as an illustration.

4.5 Summary

Choosing the right method to answer the biological question of interest is a key skill in statistical analysis. This chapter outlined examples of biological questions that `mixOmics` methods can help answer.

TABLE 4.1: Different types of analyses for different types of studies with mixOmics. The types of data and the biological questions drive the different analyses.

Name	Type of data	# of samples	Treatment/group	Methods	More details
multidrug	ABC transporters (48) gene expr. (1,429)	60 cell lines	cell lines (9)	PCA, sPCA NIPALS IPCA, sIPCA	?multidrug See Chapter 9
liver.toxicity	gene expr. (3,116) clinical measures (10)	64 rats	dose (4) time necropsy (4)	sPLS multilevel sPLS	?liver.toxicity See Chapter 10
nutrimouse	lipids (21) gene expr. (120)	40 mice	diet (5) genotype (2)	rCCA	?nutrimouse See Chapter 11
srbct	gene expr. (2,308)	63 human tumour samples	tumour type (4)	PCA PLS-DA, sPLS-DA	?srbct See Chapter 12
breast.TCGA Training	mRNA (200) miRNA (184) protein (142)	150 tumour samples	tumour subtype (3) genotype (2)	block.splsda	?breast.TCGA See Chapter 13
breast.TCGA Validation	mRNA (200) miRNA (184)	70 tumour samples	tumour subtype (3) genotype (2)	predict.splsda	?breast.TCGA See Chapter 13
vac18	gene expr. (1,000)	42 PBMC human samples	stimulation (4)	multilevel PCA multilevel sPLS-DA	?vac18 See Appendix 4.A.1 & Section 12.6
diverse.16S	OTU (1,674)	162 samples	bodysites (3)	multilevel PCA multilevel sPLS-DA	?diverse.16S See Section 12.6

The package proposes several multivariate projection-based methods to perform dimension reduction, data integration, and feature selection for high-dimensional multivariate data. The methods can be categorised into single data set analysis (e.g. PCA), two (e.g. PLS, rCCA) and multiple data set integration (e.g. multiblock PLS), and across different studies (group-PLS). Both unsupervised and supervised types of analyses are available, with sparse variants to identify the most relevant variables that explain common information between data sets, or an outcome of interest. Each of these key methods will be detailed in dedicated case studies in Part III.

4.A   Appendix: Data transformations in mixOmics

4.A.1   Multilevel decomposition

In this section, we denote $X(N \times P)$ the data matrix with N the total number of samples (or rows) in the data, M the number of experimental units (or unique subjects), and P the total number of variables or predictors. In a repeated measurement design, we assume that each experimental units undergo all or most treatments G. More details about this approach are presented in Liquet et al. (2012).

4.A.2   Mixed-effect model context

Let $\boldsymbol{X}_{sj}^k$ be the gene expression of a given gene k for subject s with treatment j. The mixed-effect model is defined by:

$$\boldsymbol{X}_{sj}^k = \boldsymbol{\mu}_j^k + \boldsymbol{\pi}_s^k + \boldsymbol{\epsilon}_{sj}^k, \qquad s = 1,\ldots,M, \quad j = 1,\ldots,G_s \tag{4.1}$$

$$= \boldsymbol{\mu}_{..}^k + \boldsymbol{\alpha}_j^k + \boldsymbol{\pi}_s^k + \boldsymbol{\epsilon}_{sj}^k \tag{4.2}$$

where for a given gene k, $\boldsymbol{\mu}_j^k$ measures the fixed effect of treatment j, which can be further decomposed into $\boldsymbol{\mu}_{..}^k$, the overall mean effect treatment, plus $\boldsymbol{\alpha}_j^k$ which is the differential effect for treatment j. The $\boldsymbol{\pi}_s^k$ are independent random variables following a normal distribution $\mathcal{N}(0,\sigma_{\pi,k}^2)$, to take into account the dependency between the repeated measures made on the same subject s, while the $\boldsymbol{\epsilon}_{sj}^k$ are independent random variables following a $\mathcal{N}(0,\sigma_{\epsilon,k}^2)$ distribution. Note that the number of treatments for each subject (G_s) may differ. We also assume that $\boldsymbol{\pi}_s^k$ and $\boldsymbol{\epsilon}_{sj}^k$ are independent.

This model is also known as *one-way unbalanced random-effects ANOVA*, where one can test for differential expression between treatment groups.

4.A.3   Split-up variation

As suggested by Westerhuis et al. (2010) and according to the mixed model (4.2), the observation $\boldsymbol{X}_{sj}^k$ can be decomposed into:

$$\boldsymbol{X}_{sj}^k = \underbrace{\boldsymbol{X}_{..}^k}_{\text{offset}} + \underbrace{(\boldsymbol{X}_{s.}^k - \boldsymbol{X}_{..}^k)}_{\text{between-subject deviation}} + \underbrace{(\boldsymbol{X}_{sj}^k - \boldsymbol{X}_{s.}^k)}_{\text{within-subject deviation}} \tag{4.3}$$

where $\boldsymbol{X}_{..}^k = \dfrac{1}{N}\sum_{j=1}^{G_s}\sum_{s=1}^{M}\boldsymbol{X}_{sj}^k$ and $\boldsymbol{X}_{s.}^k = \dfrac{1}{G_s}\sum_{j=1}^{G_s}\boldsymbol{X}_{sj}^k$. In the balanced case we have $N = M \times G$, otherwise $N = \sum_{s=1}^{M}G_s$.

The offset term $\boldsymbol{X}_{..}^k$ is an estimation of $\boldsymbol{\mu}_{..}^k$. The between-subject deviation is an estimation of $\boldsymbol{\pi}_s^k$, and the within-subject deviation is an estimation of $\boldsymbol{\alpha}_j^k + \boldsymbol{\epsilon}_{sj}^k$ which can be further decomposed as:

$$\underbrace{(\boldsymbol{X}_{sj}^k - \boldsymbol{X}_{s.}^k)}_{\text{within-subject deviation}} = \underbrace{(\boldsymbol{X}_{.j}^k - \boldsymbol{X}_{..}^k)}_{\text{Treatment effect}} + \underbrace{(\boldsymbol{X}_{sj}^k - \boldsymbol{X}_{s.}^k - \boldsymbol{X}_{.j}^k + \boldsymbol{X}_{..}^k)}_{\text{error}}$$

where $\boldsymbol{X}_{.j}^k = \dfrac{1}{N_j}\sum_{s=1}^{N_j}\boldsymbol{X}_{sj}^k$ with N_j the number of subjects undergoing treatment j. Therefore, a part of the within-subject deviation is explained by the treatment effect. According to Equation (4.3):

$$\boldsymbol{X} = \underbrace{\boldsymbol{X}_{..}}_{\text{offset term}} + \underbrace{\boldsymbol{X}_b}_{\text{between-subject deviation}} + \underbrace{\boldsymbol{X}_w}_{\text{within-subject deviation}}$$

The matrix $\boldsymbol{X}_{..}$ represents the offset term defined as $\mathbf{1}_N \boldsymbol{X}_{..}^T$ with $\mathbf{1}_N$ the $(N \times 1)$ matrix contains ones and $\boldsymbol{X}_{..}^T = (\boldsymbol{X}_{..}^1, \ldots, \boldsymbol{X}_{..}^P)$; the $\boldsymbol{X}_b$ is the between-subject matrix of size $(N \times P)$ defined by concatenating $\mathbf{1}_{G_s} \boldsymbol{X}_{bs}^T$ for each subject into $\boldsymbol{X}_b$ with $\boldsymbol{X}_{bs}^T = (\boldsymbol{X}_{s.}^1 - \boldsymbol{X}_{..}^1, \ldots, \boldsymbol{X}_{s.}^P - \boldsymbol{X}_{..}^P)$; $\boldsymbol{X}_w = \boldsymbol{X} - \boldsymbol{X}_{s.}$ is the within-subject matrix of size $(N \times P)$ with $\boldsymbol{X}_{s.}$ the matrix defined by concatenating the matrices $\mathbf{1}_{G_s} \boldsymbol{X}_{s.}^T$ for each subject into $\boldsymbol{X}_{s.}$, with $\boldsymbol{X}_{s.}^T = (\boldsymbol{X}_{s.}^1, \ldots, \boldsymbol{X}_{s.}^P)$.

The mixed-model described in Equation (4.2) can provide an analysis for repeated measurements data in an unbalanced design. In `mixOmics`, we combine the multilevel approach to our multivariate approaches:

- The multilevel step splits the different parts of the variation while taking into account the repeated measurements on each subject,
- Our multivariate methods are then applied on the within matrix $\boldsymbol{X}_w$ which includes the treatment effect,
- In the case of an integrative analysis that considers more than one data set, we apply the same procedure to the other (continuous) data sets.

These steps are either implemented directly in the function with the `multilevel` argument that indicates the ID information of each sample, or by using the external function `withinVariation()` to extract the within matrix $\boldsymbol{X}_w$ as an input in the multivariate method.

We have also extended the approach for the case of two factors (e.g. time factor, before and after vaccination, in addition to the vaccination type factor), as further detailed in Liquet et al. (2012).

4.A.4 Example of multilevel decomposition in `mixOmics`

We illustrate multilevel decomposition from the `vac18` gene expression data available in the package. The `R` code indicated below can be ignored in this chapter. Peripheral Blood Mononuclear Cells from 12 vaccinated participants were stimulated 14 weeks after vaccination with four different treatments: LIPO-5 (all the peptides included in the vaccine), GAG+ (Gag peptides included in the vaccine), GAG- (Gag peptides not included in the vaccine) and NS (no stimulation). We conduct a PCA and a multilevel PCA, then display the samples projected onto the first two components. In Figure 4.2, the samples are coloured according to the treatment type, while numbers indicate the unique individual ID.

```
library(mixOmics)

# Load data
data("vac18")

# Conduct PCA
pca.vac18 <- pca(vac18$genes, scale = TRUE)

# Sample plot
plotIndiv(pca.vac18, group = vac18$stimulation,
          ind.names = vac18$sample,
          legend = TRUE, legend.title = 'Treatment',
          title = 'PCA on VAC18 data')
```

```
# Conduct PCA with multilevel argument
multilevel.pca.vac18 <- pca(vac18$genes, multilevel = vac18$sample,
                            scale = TRUE)
# sample plot
plotIndiv(multilevel.pca.vac18, group = vac18$stimulation, pch = 16,
              legend = TRUE, legend.title = 'Treatment',
              title = 'Multilevel PCA on VAC18 data')
```

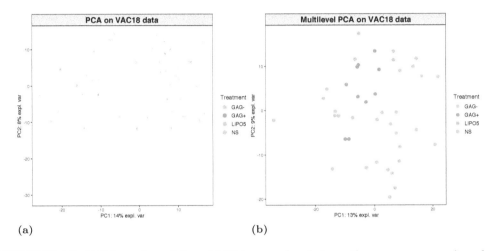

(a)   (b)

FIGURE 4.2: PCA and multilevel PCA sample plot on the gene expression data from the vac18 study. (a) The sample plot shows that samples within the same individual tend to cluster, as indicated by the individual ID, but we observe no clustering according to treatment. **(b)** After multilevel PCA, we observe some clustering according to treatment type.

We observe that samples within individual participant IDs tend to cluster. Based on this information, we then conduct a multilevel PCA (argument `multilevel` takes into account the individual ID of each sample stored in `vac18$sample`). In the sample plot, we observe a slightly improved clustering of the samples according to treatment.

4.B   Centered log ratio transformation

We assume the data are in raw count format. For microbiome data, the raw count data may result from bioinformatics pipelines such as QIIME2 (Caporaso et al., 2010) or FROGS (Escudié et al., 2017) for 16S rRNA gene amplicon data. In `mixOmics`, we usually consider the lowest taxonomy level (Operational Taxonomy Unit, OTU), but other levels can be considered, as well as other types of microbiome-derived data, such as whole genome shotgun sequencing. The data processing step is described in (Lê Cao et al., 2016). The R code is available on http://www.mixomics.org/mixMC and consists of:

1. Offset of 1 on the *whole* data matrix to manage zeros for CLR transformation (after CLR transformation, the offset zeros will still be zero values).
2. Low Count Removal: Only OTUs whose proportional counts exceeded 0.01% in at least one sample were considered for analysis. This step aims to counteract sequencing errors (Kunin et al., 2010).
3. Compositional data analysis: Total Sum Scaling (TSS) is considered as a 'normalisation' process to account for uneven sequencing depth across samples. TSS divides each OTU count by the total number of counts in each individual sample but generates compositional data expressed as proportions. Instead, one can use Centered Log Ratio transformation (CLR) directly, that is scale invariant, and addresses in a practical way the compositionality issue arising from microbiome data. CLR projects the data into a Euclidean space (Aitchison (1982); Fernandes et al. (2014); Gloor et al. (2017)). Given a vector $\boldsymbol{X}_j$ of P OTU counts for a given sample j, $j = 1, \ldots, N$, we define CLR as a log transformation of each element of the vector divided by its geometric mean $\mathcal{G}(\boldsymbol{X}_j)$:

$$\text{clr}(\boldsymbol{X}_j) = \left[\log(\frac{\boldsymbol{X}_j^i}{\mathcal{G}(\boldsymbol{X}_j)}), \ldots, \log(\frac{X_j^P}{\mathcal{G}(\boldsymbol{X}_j)}) \right] \tag{4.4}$$

where $\mathcal{G}(\boldsymbol{X}_j) = \sqrt[P]{\boldsymbol{X}_j^1 \times \boldsymbol{X}_j^2 \times \ldots \boldsymbol{X}_j^P}$.

In the context of lasso penalisation in a multivariate model, the CLR transformation has some limitations, as the denominator $\mathcal{G}(\boldsymbol{X}_j)$ (that includes the information of all OTU variables) remains present regardless of the variable selection. Such limitations are further explained in Susin et al. (2020).

In `mixOmics`, a `logratio = 'CLR'` argument is available in some of our methods. Otherwise, one can use the external function `logratio.transfo()` first to calculate the CLR data as an input in the multivariate method.

4.C   Creating dummy variables

We provide an `R` example where factor (categorical) variables are transformed into a dummy matrix `Y` using the `unmap()` function for downstream analysis:

```
library(mixOmics)
sex <- factor(c("male", "female", "female", "male", "female",
                "male", "female"))
food <- factor(c("low", "high", "medium", "high", "low",
                "medium", "high"))

sex.dummy <- unmap(sex)
colnames(sex.dummy) <- levels(sex)
food.dummy<- unmap(food)
colnames(food.dummy) <- levels(food)
```

```
Y <- data.frame(sex.dummy, food.dummy)
Y
```

```
##   female male high low medium
## 1      0    1    0   1      0
## 2      1    0    1   0      0
## 3      1    0    0   0      1
## 4      0    1    1   0      0
## 5      1    0    0   1      0
## 6      0    1    0   0      1
## 7      1    0    1   0      0
```

Part II

mixOmics under the hood

5

Projection to latent structures

In `mixOmics`, most of our methods project data onto latent structures, or components, for matrix factorisation – and hence matrix decomposition. Different methods define components differently according to a specific statistical criterion that is maximised, such as variance, covariance, or correlation. Projection to Latent Structures (PLS) is the algorithm that underpins the other methods implemented in this package. We start by describing different algorithms to solve PCA, using either singular value decomposition, or PLS.

5.1 PCA as a projection algorithm

5.1.1 Overview

In PCA, components are defined to summarise the *variance* in the data. The first principal component (PC) explains as much variance in the data as possible by optimally weighting variables across all the samples, and combining these weighted variables into a single score, in a linear manner. This score corresponds to the sample's *projection* into the space spanned by the first component.

The second PC attempts to explain some of the remaining variance present in the data that was not explained by the first component, and so on for the remaining components. Thus, the components explain a successively decreasing amount of variance. In statistical terms, we say that the components are *orthogonal*, a property which avoids any overlap of information that is extracted between them.

Components are artificial variables that aggregate the original variables. They represent a new coordinate space that is much smaller than the original, high-dimensional data space. Samples projected into this more comprehensible space enable us to better understand the major sources of variation in the data (Figure 5.1).

The PCA process through matrix decomposition (see Section 3.2) results in a number of components equal to the *rank* of the original data matrix[1]. However, in order to reduce the dimensions of the data, only the first few components explaining the largest sources of variation are retained. By doing so, we assume that the remaining components may only explain noise or irrelevant information to the biological problem at hand. Therefore, we need to keep in mind that dimension reduction results in data approximation and a loss of information but can provide deeper insight into the data by highlighting experimental and biological variation. When using matrix factorisation techniques, we often refer to *dimensions*

---

[1]The rank of a matrix is the number of variables that are uncorrelated to each other, or the number of variables that are non-redundant.

DOI: 10.1201/9781003026860-5

PCA projection

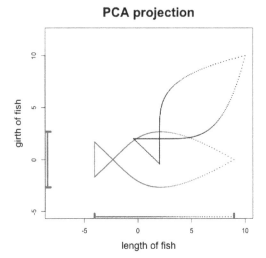

FIGURE 5.1: PCA summarises complex data into meaningful dimensions through data projection. For example, consider a fish (in black) in two dimensions. After PCA, the fish is rotated (blue), so that if we project the data points of the rotated fish onto the first dimension or component (x-axis), PC1 explains the length of the fish. Similarly, if we perform a projection of the data points onto the second dimension (y-axis), PC2 explains the girth of the fish, i.e. some of the remaining variance not captured by PC1. Thus the PCs constitute a new coordinate system through projection. The red points on the x-axis represent the 'extreme' data points after projection and illustrate that the variance of the first PC is maximised. The variance of PC2 is smaller than of PC1 and each component describes new information not available from the previous component. We could expand this example to a more complex scenario where a fish would be measured according to hundreds of parameters, and then summarised via key descriptive attributes with PCA.

as the number of components chosen to summarise the data. We discuss in more detail in subsequent chapters how to choose the number of dimensions to retain for each method.

5.1.2   Calculating the components

In this book, we will use bold font to denote either vectors or matrices and elements from these vectors or matrices. We denote X a matrix of size $(N \times P)$ with P column vectors $X^1, X^2, \ldots, X^P$ corresponding to the P variables. Each variable is weighted with a coefficient $\{a_1, a_2, \ldots, a_P\}$ that is specified in a *loading vector* $a = (a_1, a_2, \ldots, a_P)$.

A component is a *linear combination* of the original variables:

$$c = \{a_1 * X^1 + a_2 * X^2 + \ldots + a_P * X^P\}. \tag{5.1}$$

where the variable X^1, for example, represents the expression values of gene 1 across *all* samples. We assign each of the weights $(a_1, a_2, \ldots, a_P)$ to their corresponding variable vector $X^1, X^2, \ldots, X^P$. The weight in absolute value reflects the variable's importance in defining the component c, which is a vector.

5.1.3   Meaning of the loading vectors

Intuitively, if we consider the absolute values of the loading vector coefficients, a large value indicates that the variable is influential in defining the component. Conversely, a value close to zero indicates a variable with very little effect in defining the component. Also, variables with weights of the same sign and with a large absolute value are likely to be highly positively correlated, while variables with different signs but with a large absolute value are likely to be negatively correlated. We will extend that concept when visualising the correlations between variables in Section 6.2. Loading vectors differ from one dimension to another to ensure the orthogonality property of their corresponding components.

Determining the coefficients in the loading vector a is a key task, as they are used to calculate the components (Equation (5.1)). In effect, this requires solving a system of linear equations to determine eigenvalues and eigenvectors. They can be calculated using different algorithms such as Singular Value Decomposition (SVD, Golub and Reinsch (1971)) or Non-linear Iterative Partial Least Squares (NIPALS, Wold (1966)), as we detail in Section 5.2.

5.1.4   Example using the `linnerud` data in `mixOmics`

We return to the `linnerud` data where we calculated the covariance and correlation between the Exercise variables in Chapter 3.1. The amount of explained variance is output per component:

```
library(mixOmics)
data("linnerud")
data <- linnerud$exercise
pca.result <- pca(data, ncomp = 3, center = FALSE, scale = FALSE)

pca.result$sdev
```

```
##      PC1     PC2     PC3
## 181.614  34.201   3.802
```

We extract the loading vectors necessary to calculate the components[2]:

```
# Loading vectors
pca.result$loadings$X
```

```
##              PC1      PC2      PC3
## Chins    0.05692  0.02435 -0.99808
## Situps   0.88733  0.45698  0.06175
## Jumps    0.45760 -0.88915  0.00440
```

Individual 1 has the following Exercise values:

```
data[1,]
```

---

[2]For illustrative purposes, we do not center nor scale the data. The impact of doing so is detailed in Chapter 8.

```
##    Chins Situps Jumps
## 1     5    162    60
```

The *component score* of the Exercise variables for this individual is calculated as the linear combination:

$$
\begin{bmatrix} Chins & Situps & Jumps \\ 0.06 & 0.89 & 0.46 \end{bmatrix} \times \begin{bmatrix} Indiv & 1 \\ & 5 \\ & 162 \\ & 60 \end{bmatrix} = 0.06 \times 5 + 0.89 \times 162 + 0.48 \times 60 = 172.08
$$

where the loading vector values have been rounded to the second decimal. Similarly, we calculate the scores for the remaining individuals with the same variable loading weights and obtain the first principal component, shown in Figure 5.2 **(a)** in the first column. Up to some rounded value, we can see that the score of the first individual (171.49) corresponds approximately to what we have calculated above 'manually' (172.08).

When we plot principal component 1 (x-axis) versus principal component 2 (y-axis) we obtain the coordinates of each individual based on their respective linear combination of variables (Figure 5.2 **(b)**):

```
plot(pca.result$variates$X[,1], pca.result$variates$X[,2],
    xlab = 'PC1', ylab = 'PC2')
```

Let us have a closer look at the first three individuals denoted 1, 2, and 3, coloured in blue. Their coordinates on the x-axis correspond to a linear combination of the exercise value calculated as shown above for the first principal component and stored in PC1 in Figure 5.2 **(a)**. The second component is calculated similarly, based on the second loading vector (PC2 in Figure 5.2 **(b)**), as we further detail in the next section using singular value decomposition.

5.2 Singular Value Decomposition (SVD)

5.2.1 SVD algorithm

SVD is the most efficient method for performing matrix factorisation. In the particular context of PCA, it determines the loading vectors and the components from our linear combination formula in Equation (5.1) as:

$$
\boldsymbol{X} = \boldsymbol{U}\boldsymbol{\Delta}\boldsymbol{A}^{T}. \tag{5.2}
$$

In Equation (5.2), the matrix $\boldsymbol{U}$ is of size $(N \times r)$ and contains r column vectors where $r =$ rank of the matrix, each being orthogonal to each other. $\boldsymbol{\Delta}$ is a $(r \times r)$ square matrix of zero values except on the diagonal, which contains the eigenvalues $(\delta_1, \delta_2, \ldots, \delta_r)$ in decreasing order of magnitude. These values divided by $\sqrt{N-1}$ correspond to the square root of the variance explained by each of the components.

	PC1	PC2
1	171.49	20.80
2	125.18	−3.03
3	136.52	−43.36
4	110.78	15.38
5	164.82	19.58
6	109.07	8.91
7	107.46	12.56
8	129.56	21.70
9	196.62	56.19
10	338.09	−107.17
11	124.84	21.46
12	239.70	−5.97
13	239.62	5.23
14	67.30	−21.58
15	76.64	4.57
16	241.94	−10.44
17	64.91	5.29
18	241.32	34.24
19	233.91	38.28
20	117.40	12.08

(a) (b)

FIGURE 5.2: Linnerud exercise data and PCA outputs. (a) The score on PC1 and PC2 of each individual is a linear combination of the three Exercise variables weighted by their corresponding loadings. (b) Each individual resides in the space spanned by the first two principal components according to their coordinate values. These values are defined by the linear combination of their scores in the raw Exercise variables (see in (a) the values of the corresponding individuals as coloured accordingly, for example).

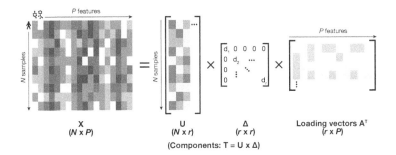

FIGURE 5.3: Singular Value Decomposition of the data matrix X into three sub-matrices. The SVD of the data matrix X with N samples in rows and P variables in columns results in its decomposition into three matrices U, Δ, and A. In matrix algebra, we call the vectors from U and A the left and right singular (eigen)vectors, respectively, and the singular values in Δ the eigenvalues from the variance-covariance matrix of X, which is $X^T X$. In fact, solving PCA consists of the eigenvalue decomposition of $X^T X$ which is a rather large matrix of size $(P \times P)$. Using SVD is a computationally efficient way of solving PCA and is adopted by most software.

TABLE 5.1: Linnerud exercise data. Principal components calculated with SVD.

PC1	PC2	PC3
171.49	20.80	5.28
125.18	−3.03	5.06
136.52	−43.36	−5.30
110.78	15.38	−5.33
164.82	19.58	−3.15
109.07	8.91	2.43
107.46	12.56	−1.58
129.56	21.70	1.91
196.62	56.19	−2.44
338.09	−107.17	−0.37
124.84	21.46	−9.39
239.70	−5.97	0.50
239.62	5.23	−0.23
67.30	−21.58	2.31
76.64	4.57	−1.53
241.94	−10.44	1.52
64.91	5.29	−0.18
241.32	34.24	3.58
233.91	38.28	−0.76
117.40	12.08	4.99

The principal components are obtained by calculating the product $T = U\Delta$. Therefore, the vectors in U can be considered as components that are not standardised with respect to the amount of variance they each explain. Finally, A is a $(P \times r)$ matrix that is transposed[3] in the equation, and contains r loading vectors in columns. Each loading vector is associated to a component. These sub-matrices are represented in Figure 5.3.

5.2.2   Example in R

Let us solve PCA using SVD 'manually' (outside the package) in R on the linnerud exercise data.

- The rank of the data matrix r is 3, and thus, we can only extract a maximum of three dimensions or components.

- The components T.comp are calculated as the product between the (left) singular vectors u from the matrix U and multiplied by the singular values d, rounded to the second decimal (Table 5.1):

```
# Components:
T.comp <- svd(data)$u  %*% diag(svd(data)$d)
```

- By plotting the first and second principal components, we obtain the coordinates of the samples projected into the space spanned by these components (Figure 5.4):

---

[3]The notation T denotes the transpose of any matrix, while the notation $'$ the transpose of any vector.

TABLE 5.2: Linnerud exercise data. Loading vectors calculated with SVD.

	Dim.1	Dim.2	Dim.3
Chins	0.06	0.02	−1.00
Situps	0.89	0.46	0.06
Jumps	0.46	−0.89	0.00

TABLE 5.3: Linnerud exercise data. Variance explained per component calculated with SVD.

Comp.1	Comp.2	Comp.3
181.6	34.2	3.8

```
plot(T.comp[,1], T.comp[,2], main = 'PCA SVD')
```

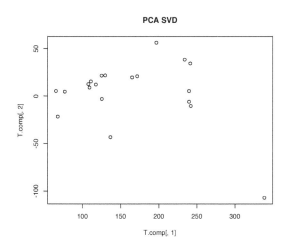

FIGURE 5.4: Sample plot on linnerud data. The graphic is obtained by plotting the first and second components calculated from the Singular Value Decomposition.

- The loading vectors from the matrix V are stored in svd(data)$v and correspond to each dimension, rounded to the second decimal (Table 5.2):

```
# Loading vectors:
svd(data)$v
```

- The amount of explained variance is given by the eigenvalues divided by $\sqrt{N-1}$, where N is the number of samples (or rows of the data matrix), as shown in Table 5.3:

```
# Variance explained by components:
svd(data)$d/sqrt(nrow(data) - 1)
```

5.2.3   Matrix approximation

From the SVD, we obtain the full matrices U, Δ, and A, each with r columns where r is the rank of the matrix X. From these matrices, we can reconstruct the full matrix X by starting from the right-hand side of Equation (5.2) and multiplying the components $U\Delta$ and the loading vectors A^T (see Figure 5.3). We will use this important property with the NIPALS algorithm described in Appendix 9.A to estimate missing values.

However, as we outlined previously, our aim is to achieve dimension reduction. Therefore, not all components and loading vectors need to be retained, but only the first few that explain the most variance or information. Thus, we can only approximate the original data matrix X after reconstruction, based on a small number of components. In statistical terms, we refer to this as *low rank approximation*, as the reconstruction is not of rank r, but lower. This approximation is the optimal lower-dimensional representation of the data that summarises most of the variance. The same underlying principles of SVD apply for the PLS-based methods in `mixOmics` and will be described in more detail in later chapters.

5.3   Non-linear Iterative Partial Least Squares (NIPALS)

We have seen how PCA can be obtained from SVD. A less direct, albeit useful algorithm to understand the methods in `mixOmics` is the NIPALS algorithm that underlies Projection to Latent Structures. We will explain in detail the important concepts of local regressions, deflation, and how missing values are handled, which will be referred to in subsequent chapters.

NIPALS was proposed by Wold (1966) to construct principal components iteratively. This algorithm forms the basis of all PLS-based methods in `mixOmics`. Understanding NIPALS will enable readers to gain a deeper insight into the algorithmic aspect of our methods. The NIPALS algorithm includes two important steps:

1. A series of **local or partial regressions** of the matrix X^T onto the component t to obtain the corresponding loading vector a, followed by a local regression of X onto the loading vector a to obtain the updated score vector t. These successive steps are repeated until the algorithm converges. Regressing the data onto a vector avoids the numerical limitation of the Ordinary Least Squares method classically used in linear regression (see Section 2.4), and is fast to calculate. We illustrate the local regressions in the next section.
2. A **deflation** step that subtracts from the original matrix X the partial reconstruction of X based on the product of the component t and loading vector a to extract the *residual* matrix of X.

Once the X matrix is deflated, the algorithm starts again based on the deflated matrix $\tilde{X}$ to define the second component and associated loading vector (i.e. the second dimension), and so on. The deflation step ensures the orthogonality between components and loading vectors: as the information explained from the previous dimension is removed, the subsequent components attempt to explain different, remaining variance.

5.3.1 NIPALS pseudo algorithm

The iterative way of solving PCA is summarised in the following algorithm (Tenenhaus, 1998; Wold, 1966):

NIPALS with no missing values

INITIALISE the first component t using SVD(X), or a column of X
  FOR EACH dimension
    UNTIL CONVERGENCE of a
    (a) Calculate loading vector $a = X^T t / t' t$ and norm a to 1
    (b) Update component $t = Xa/a'a$
    DEFLATE: current X matrix $\tilde{X} = X - ta'$ and update $X = \tilde{X}$

Notes:

- *Throughout this book, we use the notation ^{T} for the transpose of any matrix, and the notation $'$ the transpose of any vector*
- *The division by $t't$ in step (a) allows us to define $X^T t$ as a regression slope – this is particularly useful when there are missing values, as we describe below. Similarly for step (b), even if the vector t is normed.*

We focus on two important concepts here: local regressions and deflation.

5.3.2 Local regressions

Step (a) performs a local regression of each variable j onto the component t to obtain the associated weight of each variable in the loading vector a. We can therefore consider a_j as the slope of the mean squares regression line between the N data points between t and X^j, where N is the number of samples. We illustrate the local regression on the `linnerud` data by fitting a series of three linear regressions without intercept (as indicated with `-1` in the R code below) of each of the three Exercise variables onto the first principal component `T.comp[,1]` (see Figure 5.5):

```
# Plots of the fitted linear regression line with no intercept
plot(T.comp[,1], data$Chins)
abline(lm(data$Chins ~ T.comp[,1]-1))

plot(T.comp[,1], data$Situps)
abline(lm(data$Situps ~ T.comp[,1]-1))

plot(T.comp[,1], data$Jumps)
abline(lm(data$Jumps ~ T.comp[,1]-1))
```

When we extract the slopes for each variable then norm the resulting vector to 1 (using the `crossprod()` function) as mentioned in **step (a)**, we obtain the coefficients of the loading vector a as (Table 5.4):

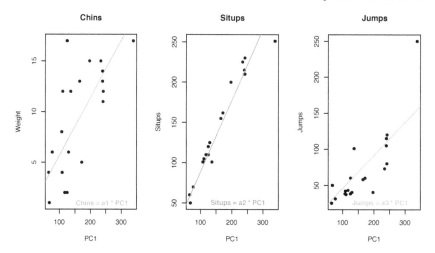

FIGURE 5.5: In the NIPALS algorithm, the loading vector coefficient for a given variable is equal to the slope of the local regression of the variable onto the principal component.

TABLE 5.4: In the NIPALS algorithm, we obtain the loading vector for the first dimension by calculating the slopes from the local regressions of each variable onto the first principal component.

	Slope
Chins	0.06
Situps	0.89
Jumps	0.46

```
# Extract each coefficient value
a1 <- coefficients(lm(data$Weight ~ T.comp[,1] -1))
a2 <- coefficients(lm(data$Waist ~ T.comp[,1] -1))
a3 <- coefficients(lm(data$Pulse ~ T.comp[,1] -1))
# Combine the coefficients into a and norm a
a <- c(a1,a2,a3)
a/crossprod(a)
```

These values correspond to the loading vector values we obtained in the SVD algorithm on the first dimension (see Section 5.2).

Similarly in **step (b)**, each element in t is the regression coefficient of X^j onto a.

The series of local regressions which are conducted, first on the component, then on the loading vector, enable us to solve a multivariate regression when the number of variables is large. By alternating between the calculation of this pair of vectors a certain number of times, we ensure the convergence of the algorithm.

5.3.3   Deflation

In the deflation step, we remove the variability already explained from X before the algorithm is repeated all over again using the deflated matrix for the subsequent iterations. We denote

$\tilde{X}$ the residual matrix of the regression of X onto t, $\tilde{X}$ is the subtraction of the rank-one approximation of X. This step ensures the orthogonality between the components obtained from all iterations. Hence, the algorithm iteratively decomposes the matrix X with a set of vectors (t, a) for each dimension.

Note:

- *Deflation is important to define the type of relationship (symmetrical, asymmetrical) between data sets in PLS models, as we further detail in Section 10.2.*

5.3.4   Missing values

Remember that missing values are declared as `NA` in `R` (see Section 2.3). NIPALS can manage values that are missing since the local regression steps **(a)** and **(b)** are based on existing values only. Using this method, we can also estimate missing values, as described in Appendix 9.A.

5.4   Other matrix factorisation methods in `mixOmics`

So far, we have introduced the concept of components and loading vectors in the context of PCA, where we seek to maximise the *variance* of the components of a single data set. The other methods available in this package differ from PCA in the statistical criterion they optimise. For example, integrative methods such as PLS or CCA seek for components that maximise the *covariance* or the *correlation* between a pair of components associated to each of the two omics data sets. Therefore, the optimisation of different statistical criteria will result in a different set of components to answer a specific biological question, as we further detail in Part III.

5.5   Summary

This Chapter established the necessary foundation to understand the meaning of components and loading vectors, and how they can be calculated either via matrix decomposition, or projection to latent structures. Using the latter, we explained how the series of iterative local regressions enabled us to circumvent the problem of the large number of variables in a multivariate regression. The local regressions are also useful to manage and estimate missing values as we will detail in Chapter 9. We will refer to similar concepts in Part III to introduce other multivariate methods for supervised analysis and data integration.

6

Visualisation for data integration

The `mixOmics` package has a strong focus on graphical representation not only to understand the relationships between omics data but also the correlations between the different types of biological variables. These plots are especially relevant when data sets contain information from different biological sources.

Each analytical method in `mixOmics` is paired with insightful and user-friendly graphical outputs to facilitate interpretation of the statistical and biological results. The various visualisation options are structured to offer insights into the resulting component scores (e.g. sample plots) and the loading vectors (e.g. feature/variable plots).

The graphical functions in `mixOmics` are based on S3 methods[1]. Consequently, all `mixOmics` methods have graphical options consistent in their parameters and outputs, and these plots are employed to:

- Assess the modelling of the multivariate methods,
- Understand the relationships modelled between the omics data sets,
- Visualise the correlation structure between the omics variables.

6.1 Sample plots using components

The package `mixOmics` employs a variety of plots to visualize how samples are projected onto the components defined by methods such as PCA or PLS-DA for one data set, PLS or CCA for the integration of two data sets, and multi-block PLS for the integration of more than two data sets. These plots allow for the visualisation of similarities between samples.

However, we must bear in mind that graphics can be misleading. For unsupervised methods, adding colours and symbols to the samples according to known phenotypes of interest may foster mis- or over- interpretation about sample clustering. It is good practice to first inspect a plot with a single colour and type of symbol to critically assess clusters of samples.

6.1.1 Example with PCA and `plotIndiv`

Here is a minimal example from the **nutrimouse** study, conducting PCA on the lipid data and using the function `plotIndiv()` (Figure 6.1 (a)):

---

[1]S3 methods enable particular functions to be generalised across R objects from different classes. For example, the S3 method `plotIndiv()` can be applied to any of the results of class PCA, CCA, PLS, etc.

DOI: 10.1201/9781003026860-6

```
library(mixOmics)
data(nutrimouse)
X <- nutrimouse$lipid
pca.lipid <- pca(X, ncomp = 2)
```

```
plotIndiv(pca.lipid, title = 'PCA on nutrimouse lipid data')
```

By default, samples are indicated with their ID number (the rownames of the original data set). The samples are projected onto the space spanned by the first two principal components, as explained in Section 5.1.

In Figure 6.1 (**b**), we added colours to distinguish between diets with the argument `group` and genotype with symbols using the argument `pch`:

```
plotIndiv(pca.lipid, group = nutrimouse$diet,
          pch = nutrimouse$genotype,
          legend = TRUE, legend.title = 'Diet',
          legend.title.pch = 'Genotype',
          title = 'PCA on nutrimouse lipid data')
```

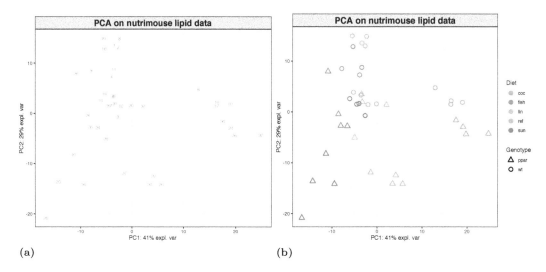

(a)   (b)

FIGURE 6.1: Sample plots from the PCA applied to the `nutrimouse` lipid data. Samples are projected into the space spanned by the first two principal components. (a) A minimal example shows the sample IDs (indicated by the rownames of the data set) and (b) A more sophisticated example with colours indicating the type of diet and symbols indicating the genotypes.

6.1.2   Sample plot for the integration of two or more data sets

When applying methods such as CCA and PLS to integrate two data sets, a pair of components and loading vectors are generated for *each* data set and each dimension – even though components are defined co-jointly so that their correlation (CCA) or covariance (PLS) across both data sets is maximised, as described later in Chapters 11 and 10. This integration is only possible between data sets with *matching samples*. The sample plots are useful here to understand the *correlation* or *covariance* structure between two data sets.

6.1.2.1 The `rep.space` argument

When integrating two data sets, the `plotIndiv()` function enables the representation of the samples into a specific projection space:

- The space spanned by the components associated to the X data set, using the argument `rep.space = 'X-variate'`,
- The space spanned by the components associated to the Y data set, using the argument `rep.space = 'Y-variate'`,
- The space spanned by the mean of the components associated to the X and Y data sets, using the argument `rep.space = 'XY-variate'`.

For example, if we integrate the `gene` and `lipid` data from the `nutrimouse` study with PLS, we obtain the following sample plot:

```
X <- nutrimouse$lipid
Y <- nutrimouse$gene
pls.nutri <- pls(X, Y, ncomp = 2)
```

```
plotIndiv(pls.nutri, group = nutrimouse$diet,
          pch = nutrimouse$genotype,  legend = TRUE,
          title = 'PLS on nutrimouse lipid and gene data')
```

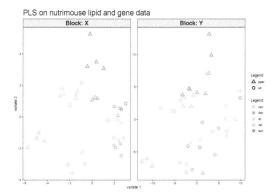

FIGURE 6.2: Sample plot from PLS regression applied to the `nutrimouse` lipid concentration (X) and gene expression (Y) data sets. This plot highlights whether samples cluster according to diet or genotype, and whether similar information can be extracted from the two data sets with this analysis. Although we are not performing a classification analysis but a multiple regression analysis, we have added colours and symbols to represent diet and phenotype as additional information.

By default, the function `plotIndiv()` shows each $X-$ and $Y-$ space separately. Each plot corresponds to the components associated to each data set, X and Y (Figure 6.2).

When specifying `rep.space = 'XY-variate'`, the average between the $X-$ and the $Y-$ components is calculated per dimension and displayed in a single plot (Figure 6.3):

```
plotIndiv(pls.nutri, group = nutrimouse$diet, pch = nutrimouse$genotype,
          rep.space = 'XY-variate',
          title = 'PLS on nutrimouse lipid and gene data')
```

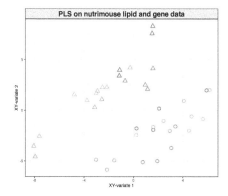

FIGURE 6.3: Example of a single sample plot between components corresponding to each data set. For a given dimension, the average between components is calculated.

The same arguments and principles apply for the integration of more than two data sets, but we use the argument `block = 'average'` that averages the components from all data sets to produce a single plot.

6.1.2.2   The `style` argument

The `plotIndiv()` function includes different styles of plot, including **ggplot2** (default), **graphics** and **lattice**. The graphics style is best for customised plots, for example to add points or text on the graph. A 3D plot that can be rotated interactively is available by specifying `style = 3d` (Figure 6.4). In the latter case, the method must include at least three components. The function will call the **rgl** library, and hence the software **XQuartz** must be installed (as explained in Chapter 8):

```
pls.nutri <- pls(X, Y, ncomp = 3)
plotIndiv(pls.nutri, group = nutrimouse$diet,
          pch = c('sphere', 'octa')[nutrimouse$genotype],
          rep.space = 'XY-variate', style = '3d')
```

6.1.2.3   The `ellipse` argument

Ellipse-like confidence regions can be plotted around specific sample groups of interest (Murdoch and Chow, 1996). In the unsupervised or regression methods, the argument `group` must be specified to indicate the samples to be included in each ellipse[2]. In the supervised

---

[2]Briefly, for each group of samples, a sample covariance matrix based on the components is calculated, along with their sample mean. Assuming a multivariate t-distribution, confidence intervals (95% level, by default) are then calculated to draw the ellipses.

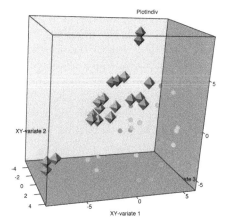

FIGURE 6.4: 3D sample plot from the PLS applied to the `nutrimouse` lipid and gene data. The plot is obtained using the argument `style = '3d'` on a PLS analysis with three components.

methods, the samples are assigned by default to the outcome of interest that is specified in the method, but other group information can be specified for the plot. The following plots in Figure 6.5 show the 95% confidence region based on the genotype of the `nutrimouse` samples in each space $X-$ and $Y-$ spanned by the components. The confidence level is set by default to 95% but can be changed using the argument `ellipse.level`.

```
plotIndiv(pls.nutri, group = nutrimouse$genotype, ellipse = TRUE,
legend = TRUE)
```

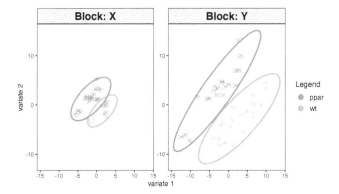

FIGURE 6.5: Sample plot with ellipses from the PLS applied to the `nutrimouse` lipid and gene data. Confidence interval ellipse plots are calculated based on the components associated to each data set for each group of interest (genotype) indicated by color. The confidence level is set to 95% here.

6.1.3 Representing paired coordinates using `plotArrow`

The arrow plot provides a complementary view of the scatter plots to represent paired coordinates, for example, from the components associated with the X and the Y data set, respectively for data integration[3]. In this plot, the samples are superimposed to visualise the agreement between data sets through the components from an integrative method. The start of the arrow indicates the location of the sample in the space spanned by the X-components, and the tip of the arrow indicates the location of the sample in the space spanned by the Y-components. Therefore, short arrows indicate agreement between data sets, while long arrows indicate disagreement (Figure 6.6). Numerically, we could calculate the correlation between components associated to their respective data set for a given dimension. A correlation close to 1 (0) would correspond to short (long) arrows. In this plot, each component is scaled first.

```
plotArrow(pls.nutri,
          group = nutrimouse$genotype,
          legend = TRUE,
          legend.title = 'Genotype')
```

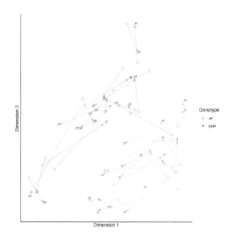

FIGURE 6.6: Arrow sample plot from the PLS applied to the `nutrimouse` lipid and gene data. Samples are coloured according to genotype. The start of each arrow indicates the location of a given sample projected onto the first two components from the gene expression data, and the tip of the arrow the location of that same sample projected onto the first two components of the lipid data. Short (long) arrows indicate strong (weak) agreement between gene expression and lipid concentration. Here, while the genotype differences are strong in each of the data sets, the agreement at the sample level using PLS is relatively weak.

This plot is useful for N-integration methods such as PLS and CCA, and was generalised for the integration of more than two data sets (multi-block methods). For example, in `block.plsda` (DIABLO), the start of the arrow indicates the centroid between component scores from all data sets for a given sample and the tips of the arrows the locations of that same sample in the different spaces spanned by the components (from each data set), as we detail in Chapter 13.

---

[3]The `plotArrow` function is an improvement of the `s.match` function from the `ade4` R package (Dray et al., 2007) to represent paired coordinates in a scatter plot.

6.2 Variable plots using components and loading vectors

In `mixOmics`, we also provide several graphical outputs to better understand the importance of the variables in defining the components and visualise their relationships.

6.2.1 Loading plots

Previous Sections 3.2 and 5.1 explained how the loading vectors define the components and thus indicate the importance of the variables during the dimension reduction process. The `plotLoadings()` function is a simple barplot representing the loading weight coefficient of each variable ranked from the most important (large weight, bottom of the plot) to the least important (small weight, top of the plot). Positive or negative signs may often reflect how a specific group of samples separate from another group of samples when projected onto a given component. When sparse methods are used, only selected variables are shown.

Different types of plots can be obtained:

- When analysing a single data set (e.g. PCA), the loading plot outputs the loading weights of the variables (Figure 6.7),
- When integrating several data sets, a barplot is output for each variable type (Figure 6.8),
- When conducting a supervised classification analysis, each variable's bar can be coloured according to whether their mean (or median) is higher (or lower) in a given group of interest (Figure 6.9).

6.2.1.1 Examples

We illustrate the representation of the loading weights on component 1 from the PCA on the lipid data (Figure 6.7):

```
plotLoadings(pca.lipid)
```

The loading weights from the lipid and gene data from PLS can be represented as (Figure 6.8):

```
plotLoadings(pls.nutri,
            subtitle = c('Lipids on Dim 1', 'Genes on Dim 1'))
```

When applying PLS-DA on the **nutrimouse** lipid data, we can colour the bars according to the genotype group in which the median of a particular lipid is maximum:

```
# PLS-DA on the lipid data
data(nutrimouse)
plsda.lipid <- plsda(nutrimouse$lipid, nutrimouse$genotype)
plotLoadings(plsda.lipid, contrib = 'max', method = 'median')
```

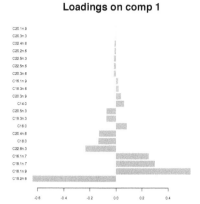

FIGURE 6.7: Loading plot from the PCA applied to the `nutrimouse` lipid data. Loading weights are represented for all lipids and are ranked from the most important (bottom) to the least important (top) to define component 1.

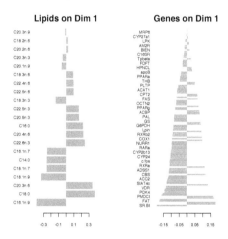

FIGURE 6.8: Loading plots from the PLS applied to the `nutrimouse` lipid and gene data. Loading weights are represented for both lipids (left) and genes (right), and are ranked from the most important (bottom) to the least important (top) on dimension 1. Lipids and genes at the bottom of the plot are likely to be highly correlated.

6.2.2   Correlation circle plots

Although relatively unknown, the correlation circle plot is an enlightening tool for data interpretation. It was primarily used for PCA outputs to visualise the contribution of each variable in defining each component, and the correlation between variables when exploring a single omics data set. We have extended these graphics to represent variables of two different types using the statistical integrative approaches CCA and PLS.

In this plot, the coordinates of the variables are obtained by calculating the correlation between each original variable and their associated component (Figure 6.10 **(a)**). Variables are centered and scaled in most of our PLS-based methods (see Section 8.1), thus, the correlation between each variable and a component can be considered as the projection of the variable on the axis defined by the component.

In a correlation circle plot, variables can be represented as vectors (Figure 6.10 **(b)**) and

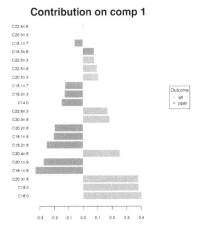

Contribution on comp 1

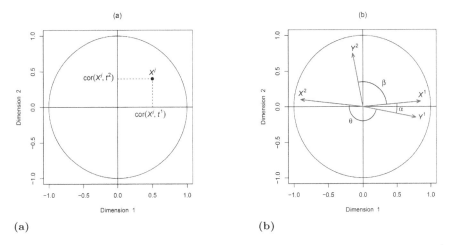

FIGURE 6.9: Loading plot from the PLS-DA applied to the nutrimouse lipid data to discriminate genotypes. Colours indicate the genotype in which the median (`method` argument) is maximum (`contrib` argument) for each lipid.

FIGURE 6.10: Correlation circle plots. (a) Coordinates of the variable X^j on the plane defined by the first two components t^1 and t^2. **(b)** The correlation between two variables is positive if the angle is sharp (e.g. α between X^1 and Y^1), negative if the angle is obtuse (θ between Y^1 and X^2), and null if the angle is right (β between X^1 and Y^2).

the relationship (correlation) between the two types of variables can be approximated by the inner product between the associated vectors[4]. In simpler terms, the nature of the correlation between two variables can be visualised through the cosine angle between two vectors representing two variables. In Figure 6.10 **(b)**, if the angle is sharp the correlation is positive (angle α between the arrows X^1 and Y^1), if the angle is obtuse the correlation is negative (angle θ between Y^1 and X^2) and if the angle is right the correlation is null (angle β between X^1 and Y^2) (González et al., 2012).

The coordinates of the variables correspond to correlation coefficient values if the variables are centered and scaled. Thus the variables are projected inside a circle of radius 1, in a space

---

[4]The inner product is defined as the product of the two vectors' lengths and their cosine angle.

spanned by a pair of components (e.g. first and second component). From the inner product definition, long arrows far from the origin indicate that a variable strongly contributes to the definition of a component. To understand this contribution, we project the tip of the arrow on the x- or the y-axis, where the x-axis (horizontal) corresponds to the first component, or dimension, and the y-axis (vertical) corresponds to the second component, or dimension (beware that the labelling on this plot can appear misleading!). When variables are close to the origin, it might be necessary to visualise the correlation circle plot in subsequent dimensions to visualise potential associations.

Correlation circle plots complement pairwise correlation approaches (González et al., 2012). As we use components as 'surrogates' to estimate the correlation between every possible pair of variables, it is computationally efficient and is also likely to avoid the identification of spurious correlations, as discussed in Section 1.4.

In `mixOmics`, we have simplified these plots to accommodate high dimensional data. Firstly, the arrows are not shown, only the tip that represents the variable. Secondly, when no variable selection is performed, the argument `cutoff` can potentially be specified to hide weaker associations.

6.2.2.1 Correlation circle plots with PCA

Let us examine the correlation circle plot from a PCA performed on the `nutrimouse` lipid data (Figure 6.11). Here we center and scale the lipids in the PCA procedure, then project the variables on the correlation circle plot either for the first two dimensions (components 1 and 2) or the first and third dimensions (components 1 and 3):

```
scale.pca.lipid <- pca(X, ncomp = 3, center = TRUE, scale = TRUE)
plotVar(scale.pca.lipid,
        title = 'Nutrimouse lipid, Components 1 and 2')
plotVar(scale.pca.lipid, comp = c(1,3),
        title = 'Nutrimouse lipid, Components 1 and 3')
```

6.2.2.2 Correlation circle plots for the integration of two data sets

Correlation circle plots can help visualise the relationships between two data sets, each containing different kinds of variables, for example, between gene expression and lipids in the `nutrimouse` data set with PLS analysis. This is possible as the variables from each data set are centered and scaled in PLS, and the covariance between components corresponding to each data set is maximised[5]. Thus, as mentioned in the previous section, the components play a surrogate role to enable comparisons between the two spaces spanned by the X- and the Y-components.

In the following, we compare the two representations, with and without a cutoff that removes the variables inside the circle of radius 0.5. Note that the argument `cutoff` is not necessary when using sparse methods that internally perform variable selection. Each colour corresponds to a type of variable, gene, or lipid (Figure 6.12):

---

[5]In fact, the correlation between components is maximised as components are scaled in the PLS procedure, as we will explain later in the PLS chapter.

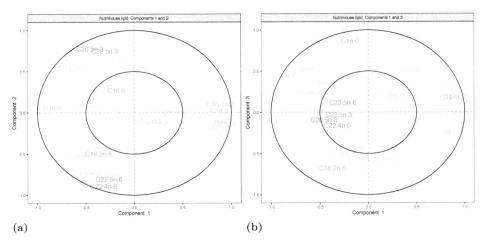

FIGURE 6.11: Correlation circle plots from the PCA applied to nutrimouse lipid data. Correlation circle plots showing the correlation structure between lipids in **(a)**, the space spanned by PC1 (x-axis, horizontal) and PC2 (y-axis, vertical), or in **(b)**, the space spanned by PC1 and PC3. Some lipids were coloured to ease interpretation. In **(a)**, we observe a strong contribution of fatty acids such as C16.1n.7, C14.0 (positive contribution) and C18.0 (negative contribution) on component 1, and C20.5n.3, C22.5n.3 (positive), C22.4n.6, C22.5n.6 (negative) on component 2. In **(b)**, we can observe the contribution of variables such as C16.0 or C18:2n.6 to component 3, which appeared more towards the inner circle of radius 0.5 in **(a)**. A further interpretation of these plots is that C16.1n.7 and C14.0 are highly positively correlated to each other, and highly negatively correlated to C18.0. Similarly for the lipids in pink, members of each pair (C20.5n.3, C22.5n.3) and (C22.4n.6, C22.5n.6) are positively correlated, while the pairs are negatively correlated to each other.

```
plotVar(pls.nutri, var.names = c(TRUE, TRUE),
        title = 'Nutrimouse, PLS no cutoff')

plotVar(pls.nutri, var.names = c(TRUE, TRUE),
        cutoff = 0.5,
        title = 'Nutrimouse, PLS with cutoff', legend = TRUE)
```

6.2.3 Biplots

In a PCA context, correlation circle plots are the premise of the *biplot* on which the sample plots are overlayed (Figure 6.13). Assuming a sufficiently small number of variables, the biplot reveals how the variables may explain particular samples: When a variable arrow points towards a group of samples, it is likely to be highly expressed in these samples. Such a conclusion should be confirmed with additional descriptive statistics.

```
biplot(scale.pca.lipid)
plotIndiv(scale.pca.lipid, group = nutrimouse$diet,
          pch = nutrimouse$genotype,
```

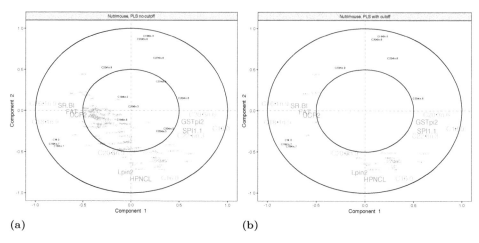

(a)   (b)

FIGURE 6.12: Correlation circle plots from the PLS applied to the `nutrimouse` lipid and gene data sets. The coordinates of each variable are obtained by calculating the correlation between each variable, and their associated component (from their respective data set). Lipids are shown in dark gray or green and genes in light gray or pink colours to ease interpretation. **(a)** includes the projection of all genes and lipids, **(b)** includes only the variables projected outside the circle of radius 0.5 by specifying the argument `cutoff`. Similar to the explanation given for Figure 6.11, we can visualise the contribution of each variable on each component, as well as the cross-correlation between variables of different types. For example, the lipids C20.3n.6, C18.0, and C22.6n.3 are highly correlated with the genes GSTpi2 and SPI1.1 with a strong contribution to the PLS component 1 (x-axis). They are not correlated to the group of lipids C16.0, C20.3n.9 and genes HPNCL, Lpin2 (to list a few) that contribute (negatively) to PLS component 2 (y-axis) but are strongly negatively correlated to C16.1n.9, C18.1n.9 and SR.BI, FAT and UCP2 on component 1.

```
legend = TRUE, legend.title = 'Diet',
legend.title.pch = 'Genotype',
title = 'PCA on nutrimouse lipid data')
```

Note:

- *The `biplot()` function is also implemented for PLS (Chapter 10) but not for other mixOmics methods currently.*

6.2.4   Relevance networks

The construction of biological networks (gene-gene, protein-protein, etc.) with direct connections within a variable set has been of considerable interest and extensively used in the literature. Here, we consider a simple, data-driven approach for modelling a network-like correlation structure *between* two data sets, using relevance networks. This concept has previously been used to study associations between pairs of variables (Butte et al., 2000). This method generates a graph with nodes representing variables, and edges representing variable associations. In `mixOmics`, since we primarily focus on data integration, the networks are *bipartite*: only pairs of variables belonging to different omics are represented.

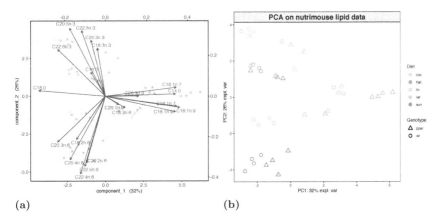

(a)   (b)

FIGURE 6.13: Biplot from the PCA applied to the nutrimouse lipid data. (a) A biplot enables visualisation of both the samples and variables on the same plot. The set of lipids that contributes positively to the definition of component 1 (as seen more clearly in the correlation circle plot in Figure 6.11 **(a)**) points towards the set of samples with a coconut oil diet, meaning that this set of lipids characterises this particular diet group. A similar interpretation can be made by overlaying the sample plot shown in **(b)** and the correlation circle plot.

The advantage of relevance networks is their ability to simultaneously represent positive and negative correlations, which are missed by methods using Euclidean distances or mutual information. Another advantage is their ability to represent variables that may belong to the same biological pathway. Most importantly for our purpose, our networks represent associations across disparate types of biological measures.

The disadvantage of relevance networks is that they may represent spurious associations when the number of variables is very large. One way to circumvent this problem would be to calculate partial correlations (i.e. the correlation between two variables while controlling for a third variable) (De La Fuente et al., 2004). In `mixOmics`, the use of variable selection within our methods, and the use of components as surrogates to estimate similarities alleviate this problem, as we describe below (see also Appendix 6.A and González et al. (2009)).

6.2.4.1 Implementation in `mixOmics`

Current relevance networks are limited by the intensive computational time required to obtain comprehensive pairwise associations when the underlying network is fully connected, i.e. when there is an edge between any pair of two types of variables. Our function `network` avoids intensive computation of Pearson correlation matrices on large data sets by calculating a *pairwise similarity matrix*. During this calculation, components are used as surrogate summary variables from the integrative approaches, such as CCA, PLS, or multi-block methods. The values in the similarity matrix represent the association between the two types of variables, e.g. variable j from the X data set, and variable k from the Y data set, denoted as the pair of vectors (X^j, Y^k). Such similarity coefficients can be seen as a robust approximation of the Pearson correlation (see González et al. (2009) for a mathematical demonstration). The similarity values between pairs of (X^j, Y^k) variables change when more dimensions are added in the integrative model. The calculation of the similarity matrix is detailed in Appendix 6.A.

When using variable selection methods (sparse methods such as `spls` or `block.splsda`), we only estimate the similarity between variables selected, thus avoiding extensive pairwise calculations. A threshold is also proposed for the other integrative methods to remove some weaker associations using the argument `cutoff`.

6.2.4.2   Example

In relevance networks, we look for *cliques* or sub-networks of subsets of variables, where the edge colours indicate the nature of the correlation (positive, negative, strong, or weak). Each of these clusters often highlights a specific correlation structure between variables. As the components are orthogonal (see Section 5.1), we can often visualise a strong correlation between variables that contribute to the same component. In Figure 6.14, we give an example of relevance networks on the `nutrimouse` data using PLS.

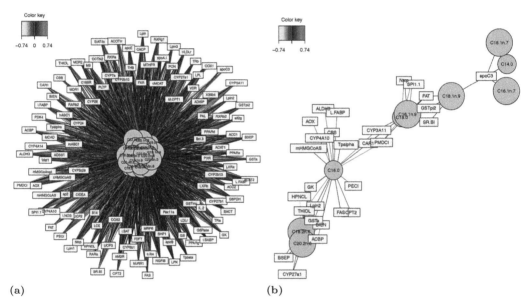

(a)   (b)

FIGURE 6.14: Relevance network from the PLS applied to the `nutrimouse` lipid and gene data sets. The correlation structure is extracted from the two data sets via the PLS components with **(a)** all variables or **(b)** variables for which the association (in absolute value) is greater than a specified cutoff of 0.55.

```
network(pls.nutri, color.node = c("orange","lightblue"))

network(pls.nutri, color.node = c("orange","lightblue"),
        cutoff = 0.55)
```

6.2.4.3   Additional arguments with `network`

We use the package `igraph` to generate the networks, which uniformly randomly places the vertices of the graph to best use the available space (Csardi et al., 2006). This means that the appearance of the network may differ in its layout when the function is rerun. To customize the network further, we advise exporting the graph to a Cytoscape file format using `write.graph()` from the `igraph` package:

```
library(igraph)
network.res <- network(pls.nutri)
write.graph(network.res$gR, file = "network.gml", format = "gml")
```

The `network()` function includes the arguments `save` and `name.save` to save the plots in formats 'jpeg', 'tiff', 'png' or 'pdf' when the RStudio window is too small. For example:

```
network.res <- network(pls.nutri, save = 'jpeg', name.save = 'mynetwork')
```

Another option is to open a new window with `X11()`.

Finally, the similarity matrix can be extracted from the network as:

```
network.res <- network(pls.nutri)
network.res$M
```

6.2.5 Clustered Image Maps (CIM)

CIM, also called 'Clustered correlation' or 'heatmaps', were first introduced to represent either the expression values of a single data set (Eisen et al., 1998; Weinstein et al., 1994, 1997), or the Pearson correlation coefficients between two matched data sets (Scherf et al., 2000). This type of representation is based on a hierarchical clustering simultaneously operating on the rows and columns of a matrix.

Dendrograms (tree diagrams) on the side of the image illustrate the clusters produced by the hierarchical clustering and indicate the similarities between samples and/or variables. The colours inside the heatmap indicate either the values of a single matrix (e.g. gene expression values) or the values of the similarity matrix when performing two data set integration. In this plot we look for 'well defined' large rectangles or squares of the same colour corresponding to long branches of the dendrograms.

For data integration methods, `cim()` is a visualisation tool that complements well the correlation circle plots and the relevance networks as we can observe clusters of subsets of variables of the same type correlated with subsets of variables of the other type. Our function takes as input the same similarity matrix as in the relevance networks when integrating two data sets (with CCA or PLS, see details in Appendix 6.A). For single omics methods (PCA, PLS-DA, and variants), the original data X are input. Similar to any hierarchical clustering method, the type of distances and linkage method must be specified for each data set space. By default, these parameters are `euclidean` distance and `complete` linkage. Our function outputs a combined number of components from the multivariate method, but specific components can also be displayed by specifying the argument `comp`.

The following code outputs a CIM for the PLS results on the `nutrimouse` study from Section 6.1, to visualise the similarities between the genes and the lipids. Two different CIM outputs are shown in Figure 6.15.

```
# On component 1:
cim(pls.nutri, xlab = "Genes", ylab = "Lipids", comp = 1)
# Default option: all components
cim(pls.nutri, xlab = "Genes", ylab = "Lipids")
```

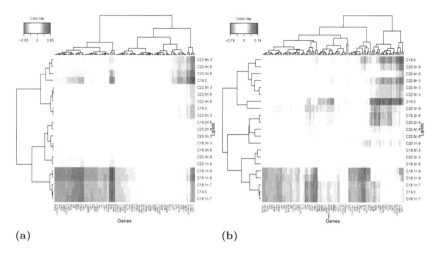

(a)   (b)

FIGURE 6.15: Clustered Image Maps from the PLS applied to the `nutrimouse` lipid and gene data to represent the correlation structure extracted from the two data sets via the PLS components. In **(a)**, the similarity input matrix is calculated on dimension 1 only, and in **(b)** the similarity input matrix is calculated across all dimensions of the model as the default option (here two dimensions). The horizontal axis shows the gene variables from X and their associated dendrogram clustering on top of the plot. The vertical axis shows the lipids from Y and their associated dendrogram clustering on the left. Red (blue) indicates a high positive (negative) similarity (similar to a correlation coefficient).

Similarly to `network()` presented above, our `cim()` function proposes the arguments `save` and `name.save` to save the plot when the RStudio window is too small. Opening a new window with `X11()` may also work.

6.2.6   Circos plots

Circos plots help to visualise associations between several types of data and is currently only applicable for the multi-block PLS-DA method `block.plsda` and `block.splsda` (Chapter 13). The association between variables is calculated using a similarity score, extended from the methods used in relevance networks and clustered image maps presented in the previous sections for more than two data sets.

The association between variables is displayed as a coloured link inside the plot to represent a positive or negative correlation above a user-specified threshold `cutoff`. Both within and between data set associations can be visualised. The variables selected by the method `block.splsda` are represented on the side of the circos plot, with side colours indicating the type of data. Optional line plots around the plot represent the expression levels in each sample group (Figure 6.16).

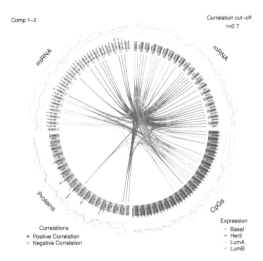

FIGURE 6.16: Circos Plot representing the associations between different types of omics variables (methylation CpGs, miRNA, mRNA, protein) across different subtypes of breast cancer (Basal, Her2, Luminal A, Luminal B) with `block.splsda` analysis. The outer lines show the expression levels or abundances of the variables for the different breast cancer subtypes. The middle ring represents the variables within the different omic types, while the inner lines represent positive (red) or negative (blue) correlations between variables. We will revisit this multi-omics example in Chapter 13.

6.3 Summary

The `mixOmics` package provides insightful and user-friendly graphical outputs to interpret multivariate analyses. By leveraging the dimension reduction outputs, the components, and the loading vectors, we can visualise the samples and the variables. The graphical functions in `mixOmics` use consistent argument calls, enabling plot customisation to obtain a full understanding of biological data.

Samples are projected into the space spanned by the components and visualised in scatterplots. The agreement extracted from the integration methods can be further understood at the sample level by visualising the samples projected into each data set component subspace or by using arrow plots.

Several graphics are proposed to visualise variables and their correlations. Loading plots describe how the variables contribute to the definition of each component, while correlation circle plots provide insightful visualisation of the correlation structure between variables, within a single molecular type, or across different types. Similarity matrices are calculated to estimate pairwise associations in an efficient manner via components to provide additional variable visualisations, such as relevance networks, clustered image maps, and circos plots for the integration of multiple data sets.

6.A   Appendix: Similarity matrix in relevance networks and CIM

In this section, we will use the following notations: X (Y) is an $N \times P$ $(N \times Q)$ data matrix. We denote X^j and Y^k the variables $j = 1, \ldots, P$ and $k = 1, \ldots, Q$ from X and Y respectively. We refer the reader to González et al. (2012) for more details on the methods presented below.

6.A.1   Pairwise variable associations for CCA

The association measure in CCA (presented in Chapter 11) is analogous to a correlation coefficient. Firstly, similar to a correlation circle output, the variables X^j and Y^k are projected onto a lower-dimensional space spanned by the components. Let $H \leq \min(P, Q)$ represent the selected dimensions to adequately account for the data association, and let $z^l = t^l + u^l$ represent the equiangular vector between the canonical variates (components) t^l and u^l $(l = 1, \ldots, H)$. Z is a matrix containing all the z^l vectors in columns. The coordinates of the variables X^j and Y^k are obtained by projecting them on the axes defined by z^l. The projection on the Z axes seems the most natural as X and Y are symmetrically analysed in CCA and their correlation is maximised in the method. Saporta (2006) showed that the Z vectors have the property to be the closest to X and Y, i.e. the sum of their squared multiple correlation coefficients with X and with Y is maximal.

Let $\tilde{\mathbf{X}}^j = (X_1^j, \ldots, X_H^j)'$ and $\tilde{\mathbf{Y}}^k = (Y_1^k, \ldots, Y_H^k)'$ the coordinates of the variable X^j and Y^k respectively on the axes defined by $z^1, \ldots, z^H$. These coordinates are obtained by computing the scalar inner product $X_l^j = \langle X^j, z^l \rangle$ and $Y_l^k = \langle Y^k, z^l \rangle$ $(l = 1, \ldots, H)$. As the variables X^j and Y^k are assumed to be of unit variance, the inner product is equal to the correlation between the variables from X (or from Y) and Z: $X_l^j = \mathrm{cor}(X^j, z^l)$ and $Y_l^k = \mathrm{cor}(Y^k, z^l)$.

Then, for any two variables X^j and Y^k, a similarity score can be calculated as follows:

$$D_k^j = \langle \tilde{\mathbf{X}}^j, \tilde{\mathbf{Y}}^k \rangle = (\tilde{\mathbf{X}}^j)' \tilde{\mathbf{Y}}^k \tag{6.1}$$

where $0 \leq |D_j^k| \leq 1$. The matrix D can be factorised as $D = \tilde{X}\tilde{Y}^T$ with $\tilde{X}$ and $\tilde{Y}$ matrices of order $(P \times H)$ and $(Q \times H)$ respectively. When $H = 2$, D is represented in the correlation circle by plotting the rows of $\tilde{X}$ and the rows of $\tilde{Y}$ as vectors in a two-dimensional Cartesian coordinate system. Therefore, the inner product of the X^j and Y^k coordinates is an approximation of their association score.

6.A.2   Pairwise variable associations for PLS

For PLS regression mode (presented in Chapter 10), the association score D_k^j between the variables X^j and Y^k can be obtained from an approximation of their correlation coefficient. Let r be the rank of the matrix X, PLS regression mode allows for the decomposition of X and Y:

$$\boldsymbol{X} = \boldsymbol{t}^1(\boldsymbol{\phi}^1)' + \boldsymbol{t}^2(\boldsymbol{\phi}^2)' + \cdots + \boldsymbol{t}^r(\boldsymbol{\phi}^r)' \tag{6.2}$$

$$\boldsymbol{Y} = \boldsymbol{t}^1(\boldsymbol{\varphi}^1)' + \boldsymbol{t}^2(\boldsymbol{\varphi}^2)' + \cdots + \boldsymbol{t}^r(\boldsymbol{\varphi}^r)' + \boldsymbol{E}^{(r)} \tag{6.3}$$

where $\boldsymbol{\phi}^l$ and $\boldsymbol{\varphi}^l$, are the regression coefficients on the variates $\boldsymbol{t}^1, \ldots, \boldsymbol{t}^r$, and $\boldsymbol{E}^{(r)}$ is the residual matrix ($l = 1, \ldots, r$). By denoting ξ_l the standard deviation of $\boldsymbol{t}^l$, using the orthogonal properties of the variates and the decompositions in Equation (6.3) we obtain $h_l^j = \mathrm{cor}(\boldsymbol{X}^j, \boldsymbol{t}^l) = \xi_l \phi_j^l$ and $g_l^k = \mathrm{cor}(\boldsymbol{Y}^k, \boldsymbol{t}^l) = \xi_l \varphi_k^l$. Let $s < r$ represent the number of components selected to adequately account for the variable association, then for any two variables $\boldsymbol{X}^j$ and $\boldsymbol{Y}^k$, the similarity score is defined by:

$$\boldsymbol{D}_k^j = \langle \boldsymbol{h}^j, \boldsymbol{g}^k \rangle = \sum_{l=1}^{s} h_l^j g_l^k = \sum_{l=1}^{s} \xi_l^2 \phi_j^l \varphi_k^l \approx \mathrm{cor}(\boldsymbol{X}^j, \boldsymbol{Y}^k), \tag{6.4}$$

where $\boldsymbol{h}^j = (h_1^j, \ldots, h_s^j)'$ and $\boldsymbol{g}^k = (g_1^k, \ldots, g_s^k)'$ are the coordinates of the variable $\boldsymbol{X}^j$ and $\boldsymbol{Y}^k$ respectively on the axes defined by $\boldsymbol{t}^1, \ldots, \boldsymbol{t}^s$. When $s = 2$, a correlation circle representation is obtained by plotting $\boldsymbol{h}^j$ and $\boldsymbol{g}^k$ as points in a two-dimensional Cartesian coordinate system.

For PLS canonical mode, the association score $\boldsymbol{D}_k^j$ is calculated by substituting $g_l^k = \mathrm{cor}(\boldsymbol{Y}^k, \boldsymbol{u}^l)$ in Equation (6.4) for $l = 1, \ldots, s$, as in this case the decomposition of $\boldsymbol{Y}$ is given by:

$$\boldsymbol{Y} = \boldsymbol{u}^1(\boldsymbol{\varphi}^1)' + \boldsymbol{u}^2(\boldsymbol{\varphi}^2)' + \cdots + \boldsymbol{u}^r(\boldsymbol{\varphi}^r)' + \boldsymbol{E}^{(r)},$$

where $\boldsymbol{\varphi}^l$ ($l = 1, \ldots, r$), are the regression coefficients on the variates $\boldsymbol{u}^1, \ldots, \boldsymbol{u}^r$.

Then,

$$\mathrm{cor}(\boldsymbol{X}^j, \boldsymbol{Y}^k) \approx \sum_{l=1}^{s} \xi_l \sigma_l \phi_j^l \varphi_k^l = \boldsymbol{D}_k^j,$$

where σ_l is the standard deviation of $\boldsymbol{u}^l$ (see further details in González et al. (2012)).

6.A.3 Constructing relevance networks and displaying CIM

The bipartite networks are inferred using the pairwise association matrix $\boldsymbol{D}$ defined in (6.1) and (6.4) for CCA and PLS results, respectively. Entry $\boldsymbol{D}_k^j$ in the matrix $\boldsymbol{D}$ represents the association score between $\boldsymbol{X}^j$ and $\boldsymbol{Y}^k$ variables. By setting a user-defined score threshold, the pairs of variables $\boldsymbol{X}^j$ and $\boldsymbol{Y}^k$ with $|\boldsymbol{D}_k^j|$ value greater than the threshold will be displayed in the Relevance Network across the s components requested by the user. By changing this threshold, the user can choose to include or exclude relationships in the Relevance Network. This option is proposed in an interactive manner in `mixOmics`.

Relevance networks for rCCA assume that the underlying network is fully connected, i.e. that there is an edge between any pair of $\boldsymbol{X}$ and $\boldsymbol{Y}$ variables. For sPLS, relevance networks only represent the variables selected by the model. In that case, $\boldsymbol{D}_k^j$ pairwise associations are calculated based on the selected variables.

Similarly, the CIM representation for CCA and PLS is based on the pairwise similarity matrix $\boldsymbol{D}$ defined (6.1) and (6.4).

7

Performance assessment in multivariate analyses

Omics studies often include a large number of variables and a small number of samples, resulting in a high risk of overfitting if the multivariate model is not rigorously assessed. In `mixOmics`, most methods can be evaluated based on different types of performance and accuracy measures. This chapter aims to provide an in-depth understanding of the technical aspects of parameter tuning and assessment in our multivariate analysis context, as we summarise in Figure 7.1. The first step consists of choosing the appropriate parameters in each method using cross-validation. The second step fits the final multivariate model. Finally, when available, the generalisability of the method can be evaluated on an independent cohort using prediction.

We describe the concepts of cross-validation, parameter tuning, the interpretation of the performance results in practice, as well as model assessment and prediction for either regression or classification. Method-specific details follow in the relevant chapters in Part III, where these concepts are applied.

This chapter intends to give the necessary background for tuning the parameters for those interested in obtaining a statistically *optimal* model – in simple terms, a model that yields the best performance, which can be useful when the biological aim is to obtain a molecular signature. Note, however, that parameter tuning in the first step can be skipped for an exploratory approach, or when the analyst chooses parameters based on biological or convenience reasons.

7.1 Main parameters to choose

In our `mixOmics` methods, two types of parameters need to be specified:

1. The number of components, or dimensions,
2. The number of variables to select on each dimension and in each data set when using sparse methods.

As we introduced in Section 3.3, the latter corresponds to choosing the lasso penalty as we describe later in Section 7.3.

We call the process of specifying these parameters a *tuning* process if the choice is based on optimising a statistical criterion. For example, we may wish to choose the number of components that explain at least 60% of the variance[1], or to choose the number of variables in a transcriptomics data set so that we achieve the best classification of samples into known

---

[1]Note that this is in a PCA context where the variance explained highly depends on the data. Setting a threshold *a priori* may not be relevant.

DOI: 10.1201/9781003026860-7

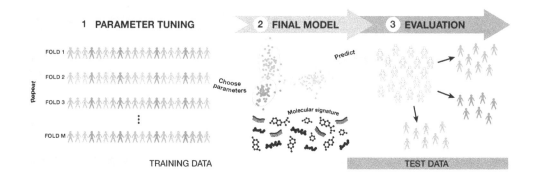

FIGURE 7.1: Overview of performance assessment. In this chapter, we describe how we can choose the optimal parameters in our multivariate methods based on performance measures. **Step 1:** Conventionally, the parameters are chosen on a training set, where artificial test sets are created using cross-validation. Cross-validation consists of dividing the data set into M subsets (or folds), fitting the model on $M-1$ subsets, and evaluating the prediction performance on the left-out subset. This process is iterated until each subset is left out once. In most cases (with the exception of leave-one-out, or leave-one-group-out) the subsets are defined randomly, and hence the process should be repeated several times before averaging the performance for each parameter value that is tuned. **Step 2:** Once the optimal parameters are chosen, the final model is applied on the full training set and the results (e.g. sample discrimination, molecular signature) are evaluated numerically and graphically. **Step 3:** If available, we can evaluate the generalisability of the trained model on an external and independent test set. When the outcome or response is already known, this step corresponds to a performance assessment task, otherwise a prediction task (here depicted as a prediction of an outcome in a classification framework).

cases and controls, compared to other variable selection sizes. We use the function `tune()` in the package.

Alternatively, when the aim is solely to explore the data, we can decide on a fixed number of components, or variables to select, regardless of any criterion to optimise. We can still evaluate the performance of our model *a posteriori*, or we can choose to not evaluate the method performance at all if our aim is purely exploratory.

7.2 Performance assessment

7.2.1 Training and testing: If we were rich

Traditionally, statistical models are fitted to a *training* data set, then applied to an external *test* study. In the case where the outcome, or phenotype of interest in the external study is *known*, we aim to assess the *generalisability* of our model. If *unknown*, we are interested in *predicting* the outcome, or phenotype of interest.

Assessing generalisability is important in omics analysis. In Section 1.4, we discussed the problem of overfitting when the number of variables is large. A statistical model that describes the training study very well may not apply well to a similar test study. We assess the performance of the method by measuring how well we can predict the outcome or the response that is already known in the test study. This gives us an estimation of performance that is as unbiased as possible. In `mixOmics`, we use the function `perf()`, as described in Section 7.3.

The prediction of an unknown outcome or phenotype is often the ultimate aim in many biomedical molecular studies for diagnosis or prognosis. We use the function `predict()` as described in Section 7.5.

Often, the number of samples in the training set is larger than the test set to ensure generalisability is met, but it depends on the type of study. For example, clinically focussed studies define the training set with an outcome or phenotype with small variability within sample groups, whereas the test set may include individuals with a larger spectrum of phenotype than in the training set.

In practice, however, we rarely have access to both a training set, and a test set, as cohort recruitment and data generation is costly. Instead, we create artificial training and testing sets within a single study, using cross-validation.

7.2.2 Cross-validation: When we are poor

Cross-validation (CV, James et al. (2013)) is a model validation technique used in statistical and machine learning to assess whether the results of an analysis can be generalised to an artificially created independent data set. It consists of dividing the data set into M subsets (or *folds*), fitting the model on $M - 1$ subsets and evaluating the prediction performance on the left-out subset. This process is iterated until each subset is left out once and the prediction performance is then averaged (Figure 7.1, **Step 1**).

There are several variants of cross-validation depending on the size of the data set, and, for a supervised classification setting, the number of samples per group:

1. M-fold CV. The data is randomly partitioned into M equal-sized subsamples. Of these M subsamples, one subsample is retained as the validation data for assessing the model, and the remaining $M - 1$ subsamples are used as training data. It is important to note that each sample is tested only once (Figure 7.1, **Step 1**).

2. Repeated M-fold CV. As samples are assigned randomly to the different folds, we repeat the CV process r times. The r performance estimations are then averaged to produce a single estimation. Repeated CV ensures that the performance evaluation does not depend on how the folds were defined.

3. Leave-one-out (LOO) CV. This process is primarily used when the number of samples is very small (<10). Each sample, in turn, is left out from the training set to be tested on so that the model is trained on $N - 1$ samples N times. As this subsampling process is not random, there is no need to repeat this process. LOO-CV is a special case of M-fold CV where $M = N$.

4. Stratified CV. In a classification setting, where samples belong to different outcome categories, we may often face an *unbalanced design* where the number of samples per category or class is not equal. It is important that the training set in the CV process is representative of the original full data set to avoid biased estimation of the performance.

Stratified CV ensures the class proportion of the samples per fold is similar to the proportions from the data.

5. Leave-One-Group-Out CV is a special case for $P-$integration when a whole study is left out. This represents a realistic training/external test set scenario, and is discussed later in Chapter 14.

7.2.2.1   How many folds should I choose?

The number of folds or subsamples M to assign samples during the CV process depends on the total number of samples N in the study. We advise choosing an M that contains at least 5–6 samples per fold (i.e. the fold test set includes at least 5–6 samples to estimate performance, $\frac{N}{M} \geq 5$). 10-fold CV is often used for large studies, whereas $M = 3$ to 5 is used for smaller studies. Finally, when $N < 10$, it is usually best to choose LOO-CV. The choice of M may result in an over- or under-estimation of performance (i.e. too optimistic, or pessimistic). Other approaches such as bootstrap and e.632+ bootstrap have been proposed as less biased alternatives (Efron and Tibshirani, 1997) but have not been implemented in `mixOmics`.

7.2.2.2   How many repeats should I choose?

During parameter tuning, the computational time increases with the number of parameters we evaluate: the number of repeats should be chosen with caution, or with a few test runs. We often advise in the final stage of evaluation to include between 50 and 100 repeats.

7.3   Performance measures

Performance can be assessed using cross-validation or using an external test set when we know the true outcome. We use different types of measures depending on the analysis. In this chapter, we specifically focus on cases where we are modelling a continuous response or phenotype (regression) or a categorical outcome (classification). However, other measures are also available for our unsupervised methods (e.g. CCA, PCA) and will be presented in subsequent chapters. The measures we briefly describe here rely on the prediction of a response or an outcome on a test sample. The concept of prediction is further described in Section 7.5.

7.3.1   Evaluation measures for regression

The evaluation of performance in a multivariate regression context is not straightforward and still requires new methodological developments. For the methods PLS and sparse PLS, we use measures of accuracy based on errors (e.g. Mean Absolute Error MAE, Mean Square Error MSE), Bias, R^2, and Q^2, as well as correlation coefficients between components and Residual Sum of Squares (RSS), see Section 10.3.

We use the following notations: y is a vector that indicates the continuous response or phenotype of each sample. We define the *error* as the difference between the predicted response of the test sample i, denoted $\hat{\boldsymbol{y}}_i$ and the observed (true) value of the response denoted $\boldsymbol{y}_i$, $i = 1, \ldots, N$. MSE and MAE are based on the errors and average the values across all CV-folds. In MSE, the errors are squared before they are averaged, while in MAE, the errors are considered in absolute values. Thus the MAE measures the average magnitude of the errors without considering their direction while the MSE tends to give a relatively high weight to large errors. The Bias is the average of the differences between the predictions $\hat{\boldsymbol{y}}_i$ and the observed responses $\boldsymbol{y}_i$, while the R^2 is defined in this context as the correlation between $\hat{\boldsymbol{y}}_i$ and $\boldsymbol{y}_i$. The Q^2 is a specific criterion developed for PLS algorithms that we mainly use to choose the number of components, as we detail further in Chapter 10 and Appendix 10.A.

The choice of the parameters is made according to the best prediction accuracy, i.e. the lowest overall error (MSE, MAE) or Bias, or the highest correlation R^2 while using cross-validation.

7.3.2 Evaluation measures for classification

For methods where the outcome is a factor, we assess the performance with the overall misclassification error rate (ER) and the Balanced Error Rate (BER). BER is appropriate for cases with an unbalanced number of samples per class as it calculates the average proportion of wrongly classified samples in each class, weighted by the number of samples in each class. Therefore, contrary to ER, BER is less biased towards majority classes during the evaluation. The choice of the parameters (described below) is made according to the best prediction accuracy, i.e. the lowest overall error rate or lowest BER.

Another evaluation is the Receiver Operating Characteristic (ROC) curve and Area Under the Curve (AUC) that are averaged over the cross-validation folds. AUC is a commonly used measure to evaluate a classifier's discriminative ability (the closer to 1, the better the prediction). It incorporates measures of sensitivity and specificity for every possible cut-off of the predicted outcomes. When the number of classes in the outcome is greater than two, we use the one-vs-all (other classes) comparison. However, as we will discuss further in Chapter 12, the AUROC may not be particularly insightful in relation to the performance evaluation of our multivariate classification methods.

7.3.3 Details of the tuning process

Whether we wish to tune the number of components, the number of variables to select, or some other parameter for more advanced N-integration methods, the tuning process, performed by the `tune()` function is similar across all methods, and is described below for different cases.

7.3.3.1 Tuning the number of components

To choose H, the optimal number of components in a multivariate method, we perform as follows:

- We evaluate several values of the parameter h. For example, we evaluate the performance of the method for a number of components varying from $h = 1$ up to 10.

- We assess each parameter value during cross-validation. Thus, if we perform 3-fold cross-validation, each of the three trained models is run for $h = 1$ to 10, and each model is tested on the left-out set: we run 10 (components) × 3 (models).

- If the cross-validation is repeated several times, we run 3 (models) × 10 (components) × the number of repeats.

The `tune()` function then outputs the performance per component by averaging the performance measure across the CV-folds and the repeats. We choose the optimal number of components H that achieves the best performance.

Figure 7.2 shows an example in a classification framework with PLS-DA, where the lowest classification error rate is obtained with five components or more (usually, we will set up for the lowest possible number of components for dimension reduction). While this decision can be made 'by eye', the `tune()` function also outputs the optimal parameter value based on one-sided t-tests to statistically assess the gain in performance when adding components to the model (see details in Rohart et al. (2017b)). In this example, according to a quantitative criterion, the optimal number of components to include is indicated as 5, but the user may choose less components.

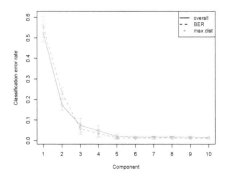

FIGURE 7.2: Performance of a PLS-DA model across ten components using three-fold cross-validation repeated ten times. The overall classification error rate and balanced error rate (y-axis) are assessed per component (x-axis) for a given prediction distance (described in Section 7.5). The standard deviation bars are calculated across the CV-folds and the repeats. The computation time was 10 seconds on a laptop for a data set with 63 samples and 2,308 genes.

7.3.3.2 Tuning the number of variables

Most packages include the lasso penalty as a direct parameter to tune (e.g. `glmnet` package from Friedman et al. (2010), `PMA` from Witten et al. (2013)). However, the lasso penalty is data-dependent, meaning that a penalty of $\lambda = 0.5$ may result in the selection of 10 genes in one data set, or 150 in another data set. In `mixOmics`, we have replaced the lasso penalty with 'the number of variables to select' as a more user-friendly, soft-thresholding algorithm. Similarly to the process described above for choosing the number of components, we illustrate how we tune the number of variables to select as a parameter value.

For example, we need to choose whether to select 5, 10, 15, 20, 25, 30, 35, 40, 45, 50, 100, 200, 300, 400, or 500 variables to achieve the best performance. We set up a grid of this parameter with these 15 values.

The performance of the model is assessed for each of the parameter values, and *for each* component. Previously we have tuned the optimal number of components to be 5. We assess the optimal number of variables across five components, using three-fold cross-validation repeated ten times. The tuning process will result in running 15 (parameters) × five (components) × 3 (folds) × 10 (repeats) = 2250 models[2]. This is run efficiently in our methods, but the user must keep in mind that the grid of parameter values needs to be carefully chosen to compromise between resolution and computational time.

The `tune()` function returns the set of variables to select that achieves the best predictive performance for each of the components that are progressively added in the model while favouring a small signature.

Figure 7.3 shows an example where we evaluate the performance of a sparse PLS-DA model for the number of variables to select per component as specified by `c(5, 10, 15, 20, 25, 30, 35, 40, 45, 50, 100, 200, 300, 400, 500)`, *per* component. The graphic indicates the optimal number of variables to select on component 1 (30), followed by component 2 (300) and 3 (45). The addition of components 4 then 5 does not seem to improve the performance further, and this is confirmed statistically in the output of the `tune()` function. The legend indicates that the process is iteratively calculated, where all parameter values are evaluated, then chosen, one component at a time, before moving to the next component. In each methods-specific chapter, we will further explain the interpretation of such outputs.

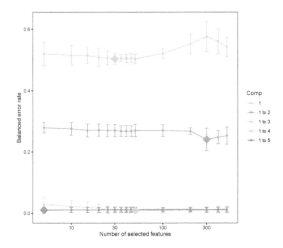

FIGURE 7.3: Performance of a sparse PLS-DA model across five components using three-fold cross-validation repeated ten times. The balanced error rate (y-axis) is assessed per component and per number of variables selected by the method (x-axis). The optimal number of variables to select per component is indicated with a diamond. The legend indicates each component's performance compounded with the model's previously tuned component. For example, the tuning for component 2 includes the optimal number of variables selected on component 1. The computation time was 1.2 minutes on a laptop for a data set with 63 samples and 2308 genes.

7.3.3.3 How do I set up the grid of parameters values?

During the tuning of the variables, consider the type of results expected for follow-up analyses. If we are interested in identifying biomarkers of disease, a small signature is preferable for knock-out wet-lab validation. If we are interested in obtaining a better understanding of the

---

[2]For a thorough tuning we would increase the number of repeats to 50.

TABLE 7.1: Example of performance output: classification error rate per class and per component allows us to understand which tumour subtype class EWS, BL, NB, or RMS is misclassified. On Component 2, NB is still difficult to classify until a third component is added to the model.

	comp1	comp2	comp3
EWS	0.01	0.06	0.03
BL	0.58	0.11	0.00
NB	0.92	0.60	0.00
RMS	0.54	0.20	0.00

pathways playing a role in a biological system, and plan to validate the signature with gene ontology, then a large signature might be preferable. We advise to first start with a coarse grid of parameter values as a preliminary run, then refine the grid as deemed appropriate.

7.4 Final model assessment

Once the parameters of the method have been chosen, either by tuning as described above, or *ad hoc*, we run a full model on the full data we have available (called 'the training set' in Figure 7.1 **Step 2**). We assess the performance of the full model on the chosen parameters using cross-validation, and, if sparse methods are used, the stability of the variables that are selected.

7.4.1 Assessment of the performance

The `perf()` function runs M-fold cross-validation using the performance measures described in Section 7.3. The difference with the tuning step is that here we solely focus on the specified parameters, and thus the process is computationally less intensive. At this stage, it is useful to examine in detail several numerical outputs. For example, for classification problems, we can look at which class is consistently misclassified as shown in Table 7.1. Different types of outputs are available and will be discussed in Chapter 12.

Another option that is not implemented in the package is to perform a permutation test (briefly introduced in Section 1.3) where sample group labels are randomly permuted several times to conclude about the significance of the differences between sample groups (Westerhuis et al., 2008).

7.4.2 Assessment of the signature

7.4.2.1 Stability

A by-product of the performance evaluation using `perf()` is that we record the variables selected on each fold across the (repeated) CV runs. The function then outputs the frequency of selection for each variable to evaluate the stability of each variable's selection, and thus

TABLE 7.2: Example of performance output: frequency of occurrence for each gene selected across the repeated CV folds. This output shows which genes are consistently selected in the signature, and can therefore be prioritised for (experimental) validation.

g879	g153	g1601	g1804	g1862	g2157	g255	g422	g742	g1662
0.7	0.67	0.67	0.67	0.67	0.67	0.67	0.67	0.67	0.63

the reproducibility of the signature when the training set is perturbed. We show an example in Table 7.2 and illustrate such outputs for most methods presented in Part III.

7.4.2.2 Graphical outputs

Sample plots and variable plots give more insight into the results. As presented in Chapter 6, the variable plots, in particular, are useful to assess the correlation of the selected features within and between data sets (correlation circle plots, clustered image maps, relevant networks, loading plots).

7.5 Prediction

When we have access to an external data set, prediction represents the ultimate aim to assess *in silico* the generalisability of the model. For example in Lee et al. (2019), the N-integration with DIABLO was further validated in a second cohort.

In Figure 7.1 **Step 3**, we illustrate the prediction principle for a classification framework, where we wish to predict the class membership of test samples. The same principle also applies in a regression framework, where we wish to predict a continuous phenotype. In this section we explain briefly how prediction is calculated. We then detail the concept of prediction distances for a classification framework that requires additional strategies.

7.5.1 Prediction of a continuous response

As introduced in Section 2.4, a multivariate model can be formulated as:

$$\boldsymbol{Y} = \boldsymbol{X}\boldsymbol{\beta} + \boldsymbol{E}, \tag{7.1}$$

where $\boldsymbol{\beta}$ is the matrix of the regression coefficients and $\boldsymbol{E}$ is the residual matrix. As explained in Chapter 5, the PLS algorithm relies on a series of local regressions to obtain the regression coefficients. We will detail in Chapter 10 how we can calculate these coefficients.

The prediction of a new set of samples from an external data set $\boldsymbol{X}_{new}$ is then:

$$\boldsymbol{Y}_{new} = \boldsymbol{X}_{new}\hat{\boldsymbol{\beta}}, \tag{7.2}$$

where $\hat{\boldsymbol{\beta}}$ is the matrix of estimated coefficients from the training step in Equation (7.1).

TABLE 7.3: Example of a prediction output for classification: the test samples are arranged in rows with their known subtypes indicated as rownames. Their prediction is calculated for each subtype class in columns as a continuous value, even if the original classes are categorical values.

	EWS	BL	NB	RMS
EWS.T2	0.772	0.027	0.067	0.134
EWS.T7	1.159	−0.056	−0.070	−0.033
EWS.C3	0.532	0.079	0.152	0.237
EWS.C4	0.450	0.096	0.181	0.273
EWS.C9	0.841	0.012	0.042	0.104
EWS.C10	0.610	0.062	0.124	0.204
BL.C5	−0.055	0.205	0.360	0.490
BL.C7	0.058	0.181	0.320	0.441
NB.C11	0.023	0.188	0.332	0.456
RMS.C2	0.101	0.171	0.305	0.423
RMS.C10	−0.045	0.203	0.357	0.486
RMS.T2	0.265	0.136	0.246	0.352
RMS.T5	0.169	0.157	0.281	0.394

Note that here the residual matrix E has vanished, as we assume we have appropriately modelled the response Y.

In multivariate analyses, the response Y can be a matrix that includes many variables, thus the prediction process is not trivial. The modelling is further complicated as we need to include several components in the model to ensure the prediction is as accurate as possible.

7.5.2 Prediction of a categorical response

The prediction framework for classification relies on the same formulation presented in Equations (7.1)–(7.2) with the exception that the response matrix Y is a dummy matrix with K columns. Each column corresponds to a binary vector and indicates whether a sample belongs to class one or two, and so forth. The method will be further described in Chapter 12. Here we outline the basic principles.

The classification algorithm relies on a regression framework for prediction. This means that the prediction is a continuous value across the K binary vectors for each sample, as we illustrate in Table 7.3, where we attempt to predict the tumour subtypes of 13 test samples. The known tumour subtypes EWS, BL, NB, RMS are indicated in each row of the predicted matrix. As the predicted matrix includes continuous values, we need to map these results to an actual class membership for easier interpretation. In mixOmics we propose several types of *distance* to assign each predicted score of a test sample to a final predicted class. The distances are implemented in all functions that rely on prediction, namely predict(), tune() and perf(). The distances are:

- max.dist, which is the simplest and most intuitive method to predict the class of a test sample. For each new individual, the class with the largest predicted score is the predicted class. In practice, this distance gives an accurate prediction for multi-class problems (Lê Cao et al., 2011).

TABLE 7.4: Example of a prediction output after applying one of the three distances (in columns): the test samples are arranged in rows, and their prediction score is assigned a class membership. The real class of each sample is indicated in the row names, illustrating several misclassifications.

	max.dist	centroid.dist	mahalanobis.dist
EWS.T2	EWS	EWS	EWS
EWS.T7	EWS	EWS	EWS
EWS.C3	EWS	EWS	EWS
EWS.C4	EWS	RMS	RMS
EWS.C9	EWS	EWS	EWS
EWS.C10	EWS	EWS	EWS
BL.C5	RMS	BL	BL
BL.C7	RMS	RMS	RMS
NB.C11	RMS	NB	NB
RMS.C2	RMS	RMS	RMS
RMS.C10	RMS	BL	BL
RMS.T2	RMS	RMS	RMS
RMS.T5	RMS	RMS	RMS

The centroids-based distances described below first calculate the centroid G_k of all the training set samples from each class k based on all components, where $k = 1, \ldots K$.

- `centroids.dist` allocates the test sample to the class that minimises the Euclidean distance between the predicted score and the centroid G_k,

- `mahalanobis.dist` measures the distance relative to the centroid using the Mahalanobis distance.

Table 7.4 shows the prediction assigned to each of the test samples from the continuous prediction matrix in Table 7.3 using each distance. We observe that some distances differ in their prediction for particular samples. For example, `max.dist` is biased towards the prediction of the RMS class. The centroid distances generally agree with Mahalanobis distances in this data set, but both misclassify some BL as RMS and vice versa.

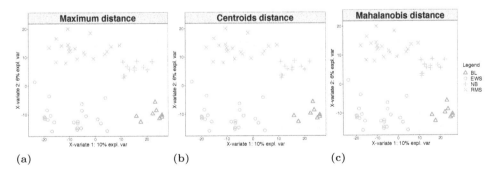

(a)   (b)   (c)

FIGURE 7.4: Visualisation of prediction distances with PLS-DA including two components, using the `background.predict()` function. (a) Maximum distance, (b) Centroids distance, and (c) Mahalanobis distance. Samples from the data set are projected onto the components. The coloured background area indicates how each sample would be predicted if they were part of a test set according to different prediction distances.

Our experience has shown that the centroids-based distances, and specifically the Mahalanobis distance, can lead to more accurate predictions than the maximum distance for complex classification problems and N-integration problems. The distances and their mathematical formulations are further detailed in Appendix 12.A and also illustrated for the N- and P-integration methods in Chapters 13 and 14.

For further illustration, the distances are visualised in a 2D sample plot using our function `background.predict()` (Appendix 12.A). Briefly, this visualisation method defines surfaces around samples that belong to the same predicted class to shade the background of the sample plot. The samples from the whole data set are represented as their projection in the space spanned by the first two components and the background colour indicates the prediction area corresponding to each class. Samples that are projected in the 'wrong' area are likely to be wrongly predicted for a given prediction distance. Figure 7.4 highlights the difference in prediction for a given distance.

7.5.3 Prediction is related to the number of components

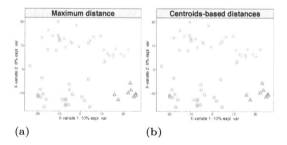

(a)   (b)

FIGURE 7.5: Visualisation of prediction accuracy when only one component is considered in PLS-DA. (a) Maximum distance, (b) Centroids or Mahalanobis distances (both give the same prediction). Contrary to Figure 7.4, the prediction area is calculated based on one component, even though samples are projected onto two components for visualisation purposes. Compared to Figure 7.4 that included two components, considering only one component in PLS-DA leads to the poor prediction of many samples.

As we have mentioned in Chapter 3 and shown in Table 7.1, the addition of several components in a multivariate model helps extract enough information from the data. Figure 7.5 illustrates the predictions we would have obtained if we had considered only one component rather than two. Of course, a compromise needs to be achieved between dimension reduction (small number of components) and performance accuracy.

7.6 Summary and roadmap of analysis

In this chapter, we discussed the important concepts of performance evaluation in supervised methods, whether on the training set to assess, in an unbiased manner, the robustness of the method's results using cross-validation, or on an external test set. Criteria and measures differ depending on the type of analysis, from regression to classification.

The main parameters to choose in multivariate methods are the number of components, and if using sparse methods, the number of variables to select per component. Depending on the aim of the analysis, one may want to lean towards an optimal solution, while for more exploratory stages, or when the number of samples is very small, the choice of parameters *ad hoc* are also acceptable and can still be evaluated (item 3 below). We can also omit the evaluation step completely if the aim of the analysis is exploratory.

For the supervised (classification and regression) `mixOmics` methods, we propose to stage the tuning process as follows:

1. Tune the number of components H using cross-validation. If using sparse methods, tune on a non-sparse model where all variables are included,
2. Sparse methods: Tune the number of variables to select on each component $h, h = 1, \ldots H$ using cross-validation. Based on the tuning results, choose specific performance criteria (e.g. error measures in PLS regression, prediction distance in PLS-DA),
3. Fit a final model on the whole training data set using the optimal parameters,
4. Assess the final performance of the model using cross-validation, and the resulting signature,
5. (optional) Assess the prediction on an external, independent test set.

For unsupervised methods where the aim might be to understand the relationships between data sets, prediction and cross-validation are not relevant. We will discuss other strategies in the subsequent chapters to choose the parameters of these methods.

Part III

mixOmics in action

8

---

mixOmics: *Get started*

In the first two parts of this book, we laid the groundwork for understanding the holistic omics data analysis paradigm and the key elements that comprise the `mixOmics` approach to multivariate analysis. This chapter describes in detail how to get started in R with `mixOmics`, from data processing to data input inside the package.

8.1 Prepare the data

As described in Section 2.3, before we start multivariate analyses with `mixOmics`, there are a few necessary pre-processing steps.

8.1.1 Normalisation

Normalisation aims to make samples more comparable and account for inherent technical biases. As normalisation is omics and platform specific, this process is not part of the package. However, the selection of a proper normalisation method is pivotal to ensure the reliability of the statistical analysis and results. Keep in mind that many normalisation methods commonly used in omics analyses have been adapted from DNA microarray or RNA-seq techniques.

In `mixOmics`, we assume the input data are normalised using techniques that are appropriate for their own type. For example, normalisation methods for sequencing count data (microbiome, RNA-seq) require pseudo counts (i.e. non zero values) for log transformation. Normalisation for RNA-seq (bulk) data considers methods such as trimmed mean of M values (TMM, Robinson and Oshlack (2010)) that estimate scale factors between samples. Microbiome data can be transformed first with centered log ratio as discussed in Section 4.1. Regarding mass spectrometry-based platforms, several methods were developed for the normalisation of label-free proteomics, including variance stabilisation normalisation (see a review from Välikangas et al. (2016)), and methods for non-targeted metabolomics are being developed (Mizuno et al., 2017). Methods for DNA methylation are numerous and include peak-based correction and quantile normalisation (see Wang et al. (2015) for a review). Finally, single cell RNA-seq normalisation methods are still in development (Luecken and Theis, 2019) and thus require keeping up to date with the literature. Note that as single cell techniques advance, we will continue improving the package so that our methods are scalable to millions of cells.

DOI: 10.1201/9781003026860-8

8.1.2   Filtering variables

While `mixOmics` methods can handle large data sets (several tens of thousands of predictors), we recommend filtering out variables that have a low variance, as they are unlikely to provide any useful information given that their expression is stable across samples. Filtering can be based on measures such as median absolute deviation or the variance calculated per variable. In microbiome data sets, we often remove variables whose total counts are below a certain threshold compared to the total counts from the data (Arumugam et al., 2011; Lê Cao et al., 2016).

Our function `nearZeroVar()` can also be used. It detects variables with very few unique values relative to the number of samples – i.e. variables with a variance close to zero. The function is called internally in our methods during the cross-validation process to avoid performance evaluation on variables with zero variance (covered in Chapter 7).

Whichever filtering method you choose, we recommend keeping at most 10,000 variables per data set to lessen the computational time during the parameter tuning process we described in Section 7.3.

8.1.3   Centering and scaling the data

As our methods are based on the maximisation of variance, correlation, or covariance, it is important to understand the effect of centering and scaling, and the default parameters (which can be changed) in each function. We start first with some basic definitions. Assuming that the variables are in columns,

- When we center (argument `center = TRUE`), we subtract from the value of each variable their sample mean,
- When we scale (argument `scale = TRUE`), we divide the value of each variable by their standard deviation.

After centering *and* scaling, each variable has a mean of 0 and a variance of 1. We call this transformation a 'z-transformation' or 'z-score':

$$z = \frac{x - m(x)}{\sqrt{v(x)}},$$

where $m(x)$ is the sample mean of the variable x and the sample variance $v(x)$ is defined as in Section 3.1 (the square root of the variance is the standard deviation). By scaling, we transform variables with different units so that they have a common scale with no physical unit.

8.1.3.1   Linnerud data

Let us have a look at the `linnerud` data, where we concatenate both Exercise and Physiological variables. As we can observe from Figure 8.1 **(a)**, each variable has a different mean and different variance. The mean is indicated by a dark symbol, and the variance or spread by a vertical line. When we center each variable in **(b)**, each variable has a mean of zero. Additionally, when we scale in **(c)**, all variables have the same variance of one. Be mindful in these plots that the scale on the y-axis is different across the plots.

We output the R code below that makes these transformations using the base `scale()` function:

```
library(mixOmics)
data(linnerud)
# Concatenate exercise and physiological variables
data <- data.frame(linnerud$exercise, linnerud$physiological)

# If we center per variable:
data.center <- scale(data, center = TRUE, scale = FALSE)
# Sample mean (rounded)
round(apply(data.center, 2, mean),2)


##   Chins Situps  Jumps Weight  Waist   Pulse
##       0      0      0      0      0       0


# Sample standard deviation (rounded)
round(apply(data.center, 2, sd),2)


##   Chins Situps  Jumps Weight  Waist   Pulse
##    5.29  62.57  51.28  24.69   3.20    7.21


# If we center and scale per variable:
data.center.scale <- scale(data, center = TRUE, scale = TRUE)
# Sample standard deviation (rounded)
round(apply(data.center.scale, 2, mean),2)


##   Chins Situps  Jumps Weight  Waist   Pulse
##       0      0      0      0      0       0


# Sample standard deviation (rounded)
round(apply(data.center.scale, 2, sd),2)


##   Chins Situps  Jumps Weight  Waist   Pulse
##       1      1      1      1      1       1
```

8.1.3.2 Example with PCA

Let us examine the effect of centering and scaling in PCA. We described in Chapter 6 the interpretation of correlation circle plots. We focus here on how the samples and variables are projected in the space spanned by the first two components.

Here is the case when we do not center, nor scale the data (Figure 8.2):

```
pca.linnerud <- pca(data, center = FALSE, scale = FALSE)
plotIndiv(pca.linnerud)
plotVar(pca.linnerud)
```

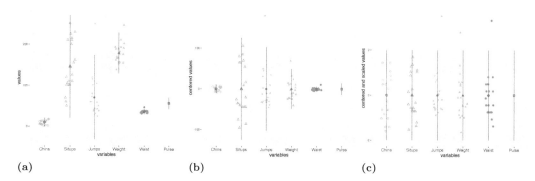

FIGURE 8.1: Linnerud exercise and physiological variables for 20 individuals.
(a) Original variables with various means and variances. **(b)** Each variable is centered and has a mean of zero. **(c)** Each variable is centered and scaled and has a mean of 0 and standard deviation of 1. Dark symbols represent the sample mean and vertical bars the sample standard deviation.

We observe that the components are not centered on zero (Figure 8.2 **(b)**). Component 1 (x-axis) highlights individuals with a large sample variance (e.g. individual 10, see also the boxplot in Figure 8.2 **(a)**) or, conversely, individuals with a small sample variance (individual 17). Component 2 (y-axis) highlights individuals with a large or small sample mean. The sample mean of the individuals clustered in the middle of the plot (e.g. individuals 1, 5, and 9) is similar to the overall mean of the data. When we do not center nor scale the data, the correlation circle plot is not valid, which explains why the variables are projected outside the circle of radius 1 (Figure 8.2 **(c)**).

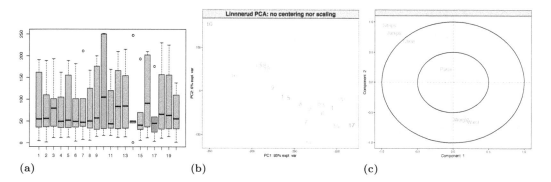

FIGURE 8.2: PCA on Linnerud exercise and physiological data when the variables are not centered nor scaled. (a) Boxplot of the values per individual. Specific individuals with high and low variance are coloured, or individuals with mean close to the overall mean. **(b)** Samples, and **(c)**, variables projected onto the space spanned by components 1 and 2. In **(c)** variables should always be centered in order to be projected on a correlation circle plot, otherwise, they may fall outside the large circle!

Using similar code as above, but changing the arguments to center but not scale the data (Figure 8.3).

```
pca.linnerud <- pca(data, center = TRUE, scale = FALSE)
```

We observe similar trends on the sample plot for individuals with high/low variance and mean (Figure 8.3 **(a)**). Component 1 explains most of the total variance (78%) and component 2 a relatively small amount (15%). The correlation circle plot in **(b)** highlights variables with a high variance (`Jumps`, `Situps`, and `Waist`, as we observed in Figure 8.1 **(b)**) on component 1 (x-axis). Remember that as the variables are not scaled, it is not appropriate to interpret the correlation between variables on this plot.

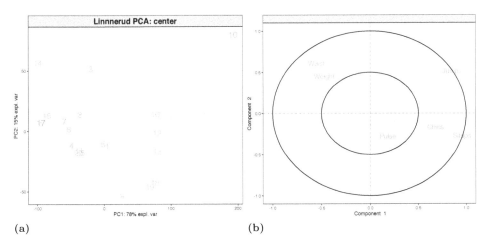

(a) (b)

FIGURE 8.3: PCA on Linnerud exercise and physiological data when the variables are centered but not scaled. (a) Samples and (b) variables projected onto the space spanned by components 1 and 2 (this plot cannot be interpreted when the variables are not scaled).

Finally, when we both center and scale the data (Figure 8.4):

```
pca.linnerud <- pca(data, center = TRUE, scale = TRUE)
```

The amount of variance explained by the first component is reduced to 54% (Figure 8.4 **(a)**). Both components are now centered on the 0 origin. Individual 10 is still driving the variance on component 1: as the data are now scaled and centered, this individual exhibits extreme values compared to the other individuals for the exercises, as shown in Figure 8.4 **(b)**, where `Chins` and `Situps` contribute to component 1 (x-axis) and are highly correlated. Variables with a small variance (`Pulse` in particular) play a more important role in the definition of both components. The variables `Waist` and `Weight` are identified as correlated in this plot.

8.1.3.3 In mixOmics

When performing PCA with `mixOmics`, data are centered by default but not scaled (argument `scale`). All PLS-related methods maximise the covariance between data sets of different scales, and hence by default, `scale = TRUE` (PLS, PLS-DA, GCCA, block and MINT methods). In these methods, the `scale` argument can be changed, depending on the question.

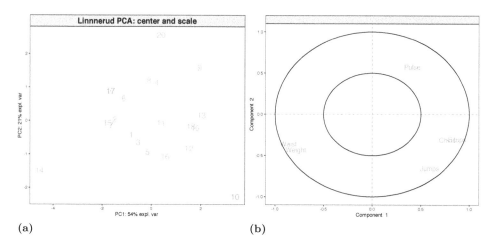

(a)   (b)

FIGURE 8.4: PCA on Linnerud exercise and physiological data when the variables are centered and scaled. (a) Samples and (b) variables projected onto the space spanned by components 1 and 2.

The CCA method maximises the correlation of the data, and therefore does not center nor scale the data.

The choice of scaling matters primarily in PCA where the aim is to understand the largest sources of variation in the data – whether biological, or technical, or unknown. Without scaling, a variable with high variance will drive the definition of the component. With scaling, a variable with low variance is considered as 'important' as any other meaningful variable, but the downside is that this variable might be noisy. Thus, it is useful to test different scenarios as we have shown in the `linnerud` example. In the PLS methods, we recommend using the default scaling as variables are of different units and scales and need to be integrated.

Note:

- *Some integrative multivariate methods such as Multiple Factor Analysis (De Tayrac et al., 2009) scale each data set according to the amount of variance-covariance they each explain. In this context, it allows a balanced representation of each data set before a global PCA is fitted on the concatenated data. In mixOmics, our methods do not apply such scaling, in part because we wish to understand the amount of variance-covariance explained by each data set, and also because we use PLS-based methods that do not concatenate data sets.*

8.1.4 Managing missing values

We have described in Section 2.3 that missing values should be declared appropriately as `NA` in `R` (or alternatively as a zero value if you believe it should correspond to a variable that is 'not present' in a particular sample). When the number of missing values is large ($>20\%$ of the total number of values in the data set), one should consider either filtering out the variables with too many missing values, or estimating (*imputing*) missing values with dedicated methods. For example in Section 5.3, we mentioned that NIPALS can be used to estimate missing values, as we will further describe in Chapter 9. Other methods can also be envisaged, for example based on machine learning approaches, such as Random Forests, neural networks, or k-nearest neighbours. However, these methods fit a model based on the independent variable (categorical or continuous) prior to missing value imputation. This means that there is a high risk of overfitting if the imputed data are then input into another supervised method for downstream statistical analysis.

8.1.5 Managing batch effects

In Section 2.2, we discussed the issue of confounders and batch effects that can be mitigated with experimental design. Here we define batch effects as unwanted variation introduced by confounding factors that are not related to any factors of interest. Batch effects might be technical (sequencing runs, platforms, protocols) or biological (cages, litter, sex). They cannot be managed directly in `mixOmics` and need to be taken into account prior to the analysis.

Several methods have been proposed to correct for batch effects, either through univariate methods such as the `removeBatchEffect` function in `limma` based on linear models (Smyth, 2005) or the Bayesian `comBat` approach (Johnson et al., 2007), or with multivariate methods (e.g. `SVA` from Leek and Storey (2007), `RUV` from Gagnon-Bartsch and Speed (2012) and PLS-DA from our recent developments Wang and Lê Cao (2020)[1]). They can be considered as another 'layer' of normalisation prior to analysis. If batch effects removal is required, be mindful that some methods apply for *known* batch effects corresponding to a covariate that was recorded (`removeBatchEffect`, `comBat`) or *unknown* (`SVA`, `RUV`). The batch effect removal process needs to be taken with caution, as these methods make strong assumptions that variables are affected systematically across batches. This assumption does not hold in microbiome studies, for example, where micro-organisms might be affected differently across small changes in the environment. In addition, the methods may remove a large part of the biological variation, especially when batch effect and treatment are strongly confounded. Thus, we need to carefully evaluate the effectiveness of batch effect removal methods, as we describe in Wang and Lê Cao (2019).

8.1.6 Data format

The input data should be numerical `R` data frames or matrices with samples (individuals) in rows and variables in columns. The matrices should be of size $(N \times P)$, where N is the number of samples or observational units (individuals) and P is the number of biological features or variables. Each data matrix should represent one type of either omics or biological feature, including clinical variables if they are measuring similar characteristics of the samples. For example, in our `linnerud` described above, we can consider a study that includes two types of variables (and hence two matrices): Exercises and Physiological variables.

N−integration: Match samples names

In the case of integrative analysis, each data set should be sample matched, i.e. the same row should correspond to the same individual across several data sets. Check that the `rownames` of the matrices are matching.

---

[1]Currently as a sister package.

P−integration: Match variable names

When combining independent studies from the same (or similar) omics platform, each data set should be variable matched, i.e. the same column should correspond to the same variable across several data sets. Check that the `colnames` of the matrices are matching.

8.2   Get ready with the software

Now that the data are ready for analysis, let's get the software installed.

8.2.1   R installation

First, install or update the latest version of R available from the CRAN website (`http://www.r-project.org`). For a user-friendly R interface, we recommend downloading and installing the desktop software Rstudio[2] after installing the R base environment.

8.2.2   Pre-requisites

Our package assumes that our users have a good working knowledge of R programming (e.g. handling data frames, performing simple calculations, and displaying simple graphical outputs), as we have covered in this chapter and in Chapter 6. If you are new to R computing, consider reading the book from De Vries and Meys (2015) or visit the site R Bloggers or Quick-R[3] to obtain a good grasp of the basics of navigating R before starting with `mixOmics`.

8.2.3   `mixOmics` download

For *Apple Mac users*, first, ensure you have installed the XQuartz software that is necessary for the `rgl` package dependency[4].

Download the latest `mixOmics` version from Bioconductor:

```
# Install BiocManager if not installed
if (!requireNamespace("BiocManager", quietly = TRUE))
    install.packages("BiocManager")

# Ensure the following returns TRUE, or follow Bioconductor guidelines
BiocManager::valid()

# Install mixOmics
BiocManager::install('mixOmics')
```

---

[2]https://www.rstudio.com/
[3]https://www.r-bloggers.com/how-to-learn-r-2/ and https://www.statmethods.net/
[4]https://www.xquartz.org

Bioconductor packages are updated every six months. Alternatively, you can install the stable but *latest* development version of the package from Github:

```
BiocManager::install("mixOmicsTeam/mixOmics")
```

Finally, if you are unable to update to the latest R and Bioconductor versions, our gitHub page also gives some instructions to install a Docker container of the stable version of the package.

8.2.4   Load the package

Check when you load the package that the latest version is installed as indicated on Bioconductor or, if installing the development version, as indicated from GitHub[5]:

```
library(mixOmics)
```

Check that there is no error when loading the package, especially for the `rgl` library (see above).

8.3   Coding practices

In this section, we provide a few tips to work with R effectively.

8.3.1   Set the working directory

The working directory is, as its name indicates, the folder where the R scripts will be saved, and where the data are stored. Set your working directory either using the `setwd()` function by specifying a specific folder where your data will be uploaded from, or in RStudio by navigating from **Session → Set Working Directory → Choose Directory . . .**

```
setwd("C:/path/to/my/data/")
```

You will need to set the working directory every time a new R session starts to upload either the data or the saved working environment and results. The easiest way is to open the working R script file, then specify **Session → Set Working Directory → Source File location**

---

[5]https://github.com/mixOmicsTeam/mixOmics/blob/master/DESCRIPTION#L4

8.3.2   Good coding practices

As we just alluded, we strongly recommend typing and saving all R command lines into an R script (in RStudio: **File → New File → Script**), rather than typing code into the console. Using an R script file will allow modification and re-run of the code at a later date. The best reproducible coding practice is to use Rmarkdown or R Notebook (in RStudio: **File → New File → R Markdown / R Notebook**[6] as a transparent and reproducible way of analysing data (Baumer and Udwin, 2015).

8.4   Upload data

8.4.1   Data sets

The examples we give in this book use data that are already part of the package. To upload your own data, check first that your working directory is set, then read your data from a `.txt` or `.csv` format, either by using **File → Import data set** in RStudio or via one of these command lines:

```
# From csv file, the data set contains header information and row names of the
# samples are in the first column
data <- read.csv("my_data.csv", row.names = 1, header = TRUE)

# From txt file, the data set contains header information and row names of the
# samples are in the first column
data <- read.table("my_data.txt", row.names = 1, header = TRUE)
```

8.4.2   Dependent variables

Often, the outcome or response vector might be stored in the data as a column, which you may need to extract. For example, if we wish to have the first column of the `linnerud` data to be set as a continuous response of interest, then we extract this column and check that it is indeed a vector:

```
Y <- data[,1]

# Check this is a vector
is.vector(Y)

# If not, you need to set Y as a vector
Y <- as.vector(Y)
```

---

[6]rmarkdown.rstudio.com/

```
# Then remove the response vector from the data for analysis
data2 <- data[, -1]
```

The other alternative is that the dependent variable is stored in another file. In that case, you can read from this file as indicated in the section above, *but make sure it is stored as a vector* (or use the `as.vector()` function).

The third alternative is to set the vector by hand. Be mindful that this may create some mistakes, for example, if you assign to a sample a wrong value! Here we create a *categorical* response (outcome) that we set as a factor:

```
# 10 cases and 10 controls in this order
Y <- c(rep('Case', 10), rep('Control', 10))

# Set as factor, option is to define the level in the order we wish
Y <- factor(Y, levels = c('Case', 'Control'))

# Check:
summary(Y)
```

8.4.3 Set up the outcome for supervised classification analyses

For classification analyses, defining factors is important in `mixOmics`. Make sure you are familiar with the base function `factor()`. Figure 8.5 shows that some methods require the dependent variable to be set as a (qualitative) factor, as indicated in orange.

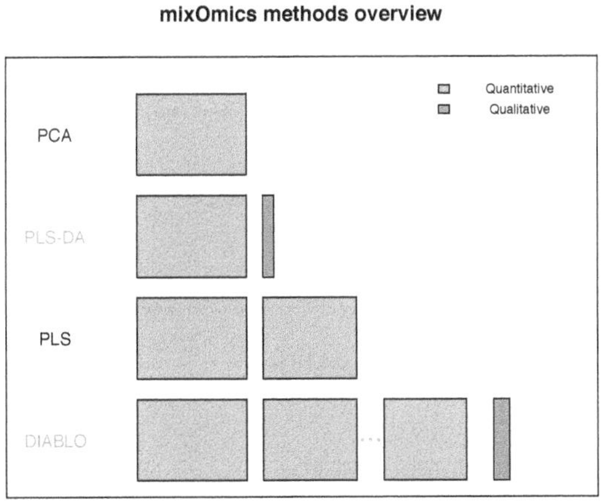

FIGURE 8.5: The choice of the method not only depends on the biological question but also on the types of dependent variables (continuous response or categorical outcome). Thus it is important that the explanatory and dependent variables (qualitative or quantitative) are well-defined in R.

8.4.4   Check data upload

Finally, check that the dimensions of the uploaded data are as expected using the `dim()` function. For example, the `linnerud` data set should contain 20 rows (e.g. samples) and 5 columns (variables) after we removed one variable to set it as a response:

```
dim(data2)
#[1] 20 5
```

If the dimensions are not as expected, the data can be examined with:

```
View(data2)    # a spreadsheet style tab opens with the entire data
head(data2)    # returns the first 6 rows of data to the console
```

For the integrative analyses of several data sets with `mixOmics`, you will need to ensure that sample **rownames** are *consistent and in the same order* from one data set to the next. Check two data sets at a time with the R base `match()` function, for example. The function will output all the rownames of the data sets `physiological` and `exercise` from the `linnerud` data that effectively match:

```
match(rownames(linnerud$physiological), rownames(linnerud$exercise))
```

```
##  [1]  1  2  3  4  5  6  7  8  9 10 11 12 13 14 15 16 17
## [18] 18 19 20
```

If this is not the case, then reorder the rows appropriately, or, if the names are not consistent, rename the row names appropriately with the `rownames()` function.

8.5   Structure of the following chapters

Part III describes in detail each method implemented in `mixOmics` as follows:

Each chapter is dedicated to one method:

1. Aim of the method,
2. Question framed biologically, and statistically,
3. Basics of the method,
4. Introduction of the case study,
5. Quick start R command lines,
6. Further options to go deeper into the analysis,
7. Additional but optional details of the methods and algorithms.

In addition to each chapter, we strongly recommend you:

- Read our glossary of terms for a refresher (Section 14.8),
- Have a look at each of the functions and their examples, e.g. `?pca`, `?plotIndiv`,
- Run the examples from the help file using the `example` function: `example(pca)`, `example(plotIndiv)`,
- Check our website `http://www.mixomics.org` that features additional tutorials and case studies,
- Keep reading this book, this is *just the beginning!*

We hope you enjoy the journey.

9

Principal Component Analysis (PCA)

9.1 Why use PCA?

Principal Component Analysis (Jolliffe, 2005) is the workhorse for linear multivariate statistical analysis. We consider PCA to be a compulsory first step for exploring individual omics data sets (e.g. transcriptomics, proteomics, metabolomics) and for identifying the largest sources of variation in the data, which may be biological or technical. After introducing PCA and useful PCA variants to perform variable selection and estimate missing values, we provide in-depth analyses of the `multidrug` study using these exploratory techniques, and indicate other extensions of PCA worth investigating. We then close with a list of Frequently Asked Questions. This chapter is an opportunity to familiarise yourself with the package and the graphical and numerical outputs, which are also used in later chapters for other types of analyses.

9.1.1 Biological questions

What are the major trends or patterns in my data? Do the samples cluster according to the biological conditions of interest? Which variables contribute the most to explaining the variance in the data?

9.1.2 Statistical point of view

PCA can be used when there is no specific outcome variable to examine, but there is an interest in identifying any trends or patterns in the data. Thus, it fits into an unsupervised framework: no information about the conditions, or class of the samples is taken into account in the method itself. For example, we may wish to understand whether any similarities between samples can be explained by known biological conditions, based on the profiling of thousands of gene expression levels.

In this chapter, we consider three types of approaches:

- Principal Component Analysis (PCA), which retains all variables. The visualisation of trends or patterns in the data allow us to understand whether the major sources of variation come from experimental bias, batch effects or observed biological differences.
- Sparse Principal Component Analysis (sPCA, Shen and Huang (2008)), which enables the selection of those variables that contribute the most to the variation highlighted by PCA in the data set. This is done by penalising the coefficient weights assigned to the variables in PCA, as introduced in Section 3.3.

- Non-linear Iterative Partial Least Squares (NIPALS, Wold (1966)) is an algorithm that solves PCA iteratively to manage or estimate missing values based on local regression (see Section 5.3).

9.2 Principle

9.2.1 PCA

The principles of PCA were detailed in Section 5.1 and are summarised here:

- PCA reduces the dimensionality of the data whilst retaining as much information (*variance*) as possible.
- A principal component is defined as a linear combination of the original variables that explains the greatest amount of variation in the data.
- The amount of variance explained by each subsequent principal component decreases, and principal components are orthogonal (uncorrelated) to each other. Thus, each principal component explains unique aspects of the variance in the data.
- Dimension reduction is achieved by projecting the data into the space spanned by the principal components: Each sample is assigned a score on each principal component as a new coordinate – this score is a linear combination of the weighted original variables.
- The weights of each of the original variables are stored in the so-called *loading vectors*, and each loading vector is associated to a given principal component.
- PCA can be understood as a matrix decomposition technique (Figure 9.1).
- The principal components are obtained by calculating the eigenvectors/eigenvalues of the variance-covariance matrix, often via singular value decomposition when the number of variables is very large (see Section 5.2).

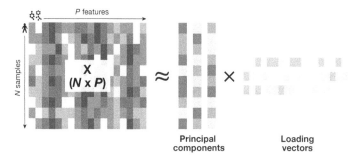

FIGURE 9.1: Schematic view of PCA matrix decomposition, where the data (centered or scaled) are decomposed into a set of principal components and loading vectors.

We use the following notations: $\boldsymbol{X}$ is an $(N \times P)$ data matrix, with N denoting the number of samples (individuals), and P the number of variables in $\boldsymbol{X}$.

The PCA objective function for the first dimension is:

$$\underset{\|\boldsymbol{a}\|=1}{\operatorname{argmax}} \operatorname{var}(\boldsymbol{X}\boldsymbol{a}) \tag{9.1}$$

where a is the first $P-$dimensional loading vector associated to the first principal component $t = Xa$, under the constraints that a is of unit (norm) 1.

We can expand the first principal component t as a linear combination of the original variables:

$$t = a_1 X^1 + a_2 X^2 + \cdots + a_P X^P$$

where $\{X^1, \ldots, X^p\}$ are the P variable profiles in the data matrix X. The first principal component t has maximal variance and $\{a_1, \ldots, a_P\}$ are the weights associated to each variable in the linear combination.

We can also write the objective function for all dimensions h (all principal components) of PCA, as

$$\underset{\|a^h\|=1,\ a^{h'}a^k=0,\ h<k}{\text{argmax}} \quad \text{var}(X^{(h-1)}a^h) \tag{9.2}$$

where $X^{(h-1)}$ are the residual matrices for dimensions $h - 1$ ($h = 1, \ldots, r$, where r is the rank of the matrix), and each loading vector a^h is orthogonal to any other previous loading vectors a^k. The residual matrix is calculated during the deflation step (see Section 5.3).

9.2.2   Sparse PCA

In `mixOmics` we have implemented the sPCA from Shen and Huang (2008) which is based on singular value decomposition. *Sparsity* is achieved via lasso penalisation (introduced in Section 3.3) on the loading vectors. Other variants of sPCA are presented in Appendix 9.A.

Sparse PCA uses the low rank approximation property of the SVD and its close link with least squares regression to calculate components and select variables (see Section 5.2). In particular, when PCA is solved with the NIPALS algorithm introduced in Section 5.3, the loading vector a for each variable is obtained by performing a local regression of X on the principal component t. Sparse PCA performs this regression step, and in addition, applies a lasso penalty on each loading vector a^h to obtain **sparse** loading vectors (Figure 9.2). Thus, each principal component is defined by the linear combination of the selected variables only. These variables are deemed most influential for defining the principal component according to the optimisation criterion. In `mixOmics`, the lasso is solved using soft-thresholding.

Notes:

- *In sparse PCA, the principal components are not guaranteed to be orthogonal. We adopt the approach of Shen and Huang (2008) to estimate the explained variance in the case where the sparse loading vectors (and principal components) are not orthogonal. The data are projected into the space spanned by the first loading vectors and the variance explained is then adjusted for potential correlations between principal components. In practice, the loading vectors tend to be orthogonal when the data are centered and scaled in the spca() function.*

- *Variables selected in a given dimension h are unlikely to be selected on another dimension because of the orthogonality property of the principal components, unless too many variables are chosen to be selected. However, variables are never discarded in the current data $X^{(h-1)}$.*

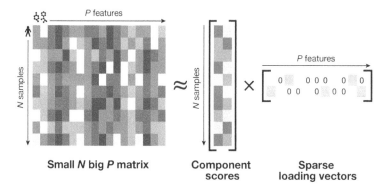

FIGURE 9.2: Schematic view of the sparse PCA method, where loading vectors are penalised using lasso so that several coefficients are shrunk to 0. As implemented in the NIPALS iterative algorithm, each principal component is then defined as a linear combination of the selected variables only (with non-zero weights).

9.3   Input arguments

9.3.1   Center or scale the data?

We discussed in Section 8.1, the importance of centering and/or scaling the variables. In PCA, variables are usually centered (`center = TRUE` by default in our function), and sometimes scaled (`scale = TRUE`). Scaling variables is recommended for the following scenarios:

- When the variance is not homogeneous across variables,

- When we wish to interpret the correlation structure of the variables and their contribution to define each component in the correlation circle plot (`plotVar()`),

- When the orthogonality between loading vectors (and thus principal components) is lost in sPCA (we found that scaling can help in obtaining orthogonal, sparse loading vectors).

9.3.2   Number of components (choice of dimensions)

To summarise the data into fewer underlying dimensions and reduce its complexity, we need to choose the number of principal components to retain. There are no clear guidelines for this choice, as it depends on the data and level of noise. However, we can visualise the proportion of total variance explained per principal component, in a 'screeplot' (barplot). By the definition of PCA, the eigenvalues are ranked from largest to smallest. We can then choose the minimal number of principal components with the largest eigenvalues before the decrease of the eigenvalues appears to stabilise (i.e. when the 'elbow' appears). We can also examine the cumulative proportion of explained variance.

Other criterion include clarity of the final graphical output (as shown in the following example in Section 9.5, bearing in mind that visualisation becomes difficult above three dimensions).

Notes:

- *The percentage of explained variance is highly dependent on the size of the data and the correlation structure between variables.*

- *The percentage of explained variance can be adjusted in sparse PCA to account for non-orthogonality between components, as proposed by Zou et al. (2006) (see also Section 9.B).*

- *Here we have only discussed empirical criteria for choosing dimensions. Some theoretical criteria have been proposed but only apply to unscaled data with Gaussian distributions. For scaled data, the Kaiser criterion commonly used in factor analysis proposes to only retain principal components with an eigenvalue >1 since the first few principal components should explain greater variance than the original variables' variance. However, this suggests using a set threshold. Karlis et al. (2003) attempted to refine this threshold based on N and P values.*

9.3.3 Number of variables to select in sPCA

Criteria to choose the number of variables to select in sPCA vary across methods and packages, as we detail in Appendix 9.A. In `mixOmics`, we propose a simple tuning strategy based on cross-validation introduced in Section 7.2:

- On the training set X_{train}, we obtain the sparse loading vector a_{train}, then calculate the predicted components based on the left-out (test) set $t_{\text{test}} = X_{\text{test}} a_{\text{train}}$.
- We then compare the predicted component values t_{test} to the actual component values obtained from performing sPCA on the whole data set X but by considering the values of the test samples only.

We run this algorithm for a grid of specified number of variables to select, then choose the optimal number of variables that maximise the correlation between the predicted and actual component values. By considering such criterion, we are seeking for a variable selection in sPCA that results in similar component values in a test set.

We illustrate this tuning strategy in Section 9.5, where we conduct repeated cross-validation. The pseudo code and more details are given in Appendix 9.A.

9.4 Key outputs

The most well-known output from PCA is the sample plot, where samples are projected into a reduced space spanned by the principal components. Other insightful graphical outputs are also available and are illustrated in the following case study.

- Sample plots enable us to visualise sample clusters and check whether those clusters agree with any biological variation introduced in the experiment (e.g. treatment).
- Correlation Circle plots (Section 6.2) project the variables into the space spanned by the principal components. Such graphics help visualise the contribution of each variable in

defining each component, as well as, the correlations between variables. For the latter, data should be scaled.

- Biplots (Section 6.2) enable a deeper understanding of the relationship between clusters of samples, and clusters of variables. This graphic is particularly insightful for small data sets.

- Numerical outputs include the proportion of explained variance per component, loading vectors that indicate the importance of each variable to define each principal component, and for sPCA, a list of variables selected on each component.

9.5   Case study: Multidrug

To illustrate PCA and is variants, we will analyse the `multidrug` case study available in the package. This pharmacogenomic study investigates the patterns of drug activity in cancer cell lines (Szakács et al., 2004). These cell lines come from the NCI-60 Human Tumor Cell Lines[1] established by the Developmental Therapeutics Program of the National Cancer Institute (NCI) to screen for the toxicity of chemical compound repositories in diverse cancer cell lines. NCI-60 includes cell lines derived from cancers of colorectal (7 cell lines), renal (8), ovarian (6), breast (8), prostate (2), lung (9), and central nervous system origin (6), as well as leukemia (6) and melanoma (8).

Two separate data sets (representing two types of measurements) on the same NCI-60 cancer cell lines are available in `multidrug` (see also `?multidrug`):

- `$ABC.trans`: Contains the expression of 48 human ABC transporters measured by quantitative real-time PCR (RT-PCR) for each cell line.

- `$compound`: Contains the activity of 1,429 drugs expressed as GI50, which is the drug concentration that induces 50% inhibition of cellular growth for the tested cell line.

Additional information will also be used in the outputs:

- `$comp.name`: The names of the 1,429 compounds.

- `$cell.line`: Information on the cell line names (`$Sample`) and the cell line types (`$Class`).

In this Chapter, we illustrate PCA performed on the human ABC transporters `ABC.trans`, and sparse PCA and the estimation of missing values with NIPALS on the compound data `compound`. To upload your own data, refer to Section 8.4.

9.5.1   Load the data

The input data matrix X is of size N samples in rows and P variables (e.g. genes) in columns. We start with the `ABC.trans` data.

---

[1]https://dtp.cancer.gov/discovery_development/nci-60/

```
library(mixOmics)
data(multidrug)
X <- multidrug$ABC.trans
dim(X) # Check dimensions of data
```

```
## [1] 60 48
```

We count the number of missing values, as this will tell us whether PCA will be solved using SVD (no missing values) or iterative NIPALS (with missing values) internally in the function pca().

```
sum(is.na(X))   # Number of NAs
```

```
## [1] 1
```

Since we have one missing value, the iterative NIPALS will be called inside pca().

9.5.2   Quick start

Here are the key functions for a quick start, using the functions pca(), spca() and the graphic visualisations plotIndiv() and plotVar().

9.5.2.1   PCA

```
result.pca.multi <- pca(X)      # 1 Run the method
plotIndiv(result.pca.multi)     # 2 Plot the samples
plotVar(result.pca.multi)       # 3 Plot the variables
```

When using minimal code, it is important to know what the default arguments are. For example here (see ?pca):

- ncomp = 2: The first two principal components are calculated and are used for graphical outputs,
- center = TRUE: Data are centered (mean = 0 for each variable),
- scale = FALSE: Data are not scaled. If scale = TRUE is employed, this standardises each variable (variance = 1).

9.5.2.2   sparse PCA

```
result.spca.multi <- spca(X, keepX = c(10, 20), ncomp = 2)
plotIndiv(result.spca.multi)
plotVar(result.spca.multi)
```

In the spca() function, we specify the number of variables keepX to select on each component (here 2 components). The default values are (see ?spca):

- `center = TRUE`: data are centered (mean = 0 for each variable),
- `scale = TRUE`: data are also scaled, as it helps to obtain orthogonal principal components.

Further interpretation is detailed below.

9.5.3   Example: PCA

Contrary to the minimal code example, here we choose to also scale the variables for the reasons detailed in Section 9.3.

9.5.3.1   Choose the number of components

The function `tune.pca()` calculates the cumulative proportion of explained variance for a large number of principal components (here we set `ncomp = 10`). A screeplot of the proportion of explained variance relative to the total amount of variance in the data for each principal component is output (Figure 9.3):

```
tune.pca.multi <- tune.pca(X, ncomp = 10, scale = TRUE)
plot(tune.pca.multi)
```

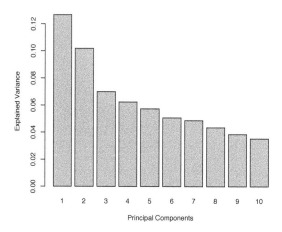

FIGURE 9.3: Screeplot from the PCA performed on the `ABC.trans` data: Amount of explained variance for each principal component on the ABC transporter data.

```
# tune.pca.multidrug$cum.var        # Outputs cumulative proportion of variance
```

From the numerical output (not shown here), we observe that the first two principal components explain 22.87% of the total variance, and the first three principal components explain 29.88% of the total variance. The rule of thumb for choosing the number of components is not so much to set a hard threshold based on the cumulative proportion of explained variance (as this is data-dependent), but to observe when a drop, or elbow, appears on the screeplot. The elbow indicates that the remaining variance is spread over many principal components and is not relevant in obtaining a low-dimensional 'snapshot' of the data. This is an empirical way of choosing the number of principal components to retain in the analysis. In this specific example we could choose between two to three components for the final

PCA, however these criteria are highly subjective and the reader must keep in mind that visualisation becomes difficult above three dimensions.

9.5.3.2 PCA with fewer components

Based on the preliminary analysis above, we run a PCA with three components. Here we show additional input, such as whether to center or scale the variables.

```
final.pca.multi <- pca(X, ncomp = 3, center = TRUE, scale = TRUE)
# final.pca.multi  # Lists possible outputs
```

The output is similar to the tuning step above. Here the total variance in the data is:

```
final.pca.multi$var.tot
```

```
## [1] 47.98
```

By summing the variance explained from all possible components, we would achieve the same amount of explained variance. The proportion of explained variance per component is:

```
final.pca.multi$prop_expl_var$X
```

```
##      PC1      PC2      PC3
## 0.12678 0.10195 0.07012
```

The cumulative proportion of variance explained can also be extracted (as displayed in Figure 9.3):

```
final.pca.multi$cum.var
```

```
##     PC1     PC2     PC3
## 0.1268 0.2287 0.2988
```

9.5.3.3 Identify the informative variables

To calculate components, we use the variable coefficient weights indicated in the loading vectors. Therefore, the absolute value of the coefficients in the loading vectors inform us about the importance of each variable in contributing to the definition of each component. We can extract this information through the `selectVar()` function which ranks the most important variables in decreasing order according to their absolute loading weight value for each principal component.

```
# Top variables on the first component only:
head(selectVar(final.pca.multi, comp = 1)$value)
```

```
##        value.var
## ABCE1     0.3242
## ABCD3     0.2648
```

```
## ABCF3    0.2613
## ABCA8   -0.2609
## ABCB7    0.2494
## ABCF1    0.2424
```

Note:

- *Here the variables are not selected (all are included), but ranked according to their importance in defining each component.*

9.5.3.4   Sample plots

We project the samples into the space spanned by the principal components to visualise how the samples cluster and assess for biological or technical variation in the data (see Section 6.1). We colour the samples according to the cell line information available in `multidrug$cell.line$Class` by specifying the argument `group` (Figure 9.4).

```
plotIndiv(final.pca.multi,
          comp = c(1, 2),     # Specify components to plot
          ind.names = TRUE,  # Show row names of samples
          group = multidrug$cell.line$Class,
          title = 'ABC transporters, PCA comp 1 - 2',
          legend = TRUE, legend.title = 'Cell line')
```

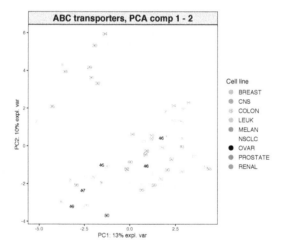

FIGURE 9.4: Sample plot from the PCA performed on the ABC.trans data. Samples are projected into the space spanned by the first two principal components, and coloured according to cell line type. Numbers indicate the rownames of the data.

Because we have run PCA on three components, we can examine the third component, either by plotting the samples onto the principal components 1 and 3 (PC1 and PC3) in the code above (`comp = c(1, 3)`) or by using the 3D interactive plot (code shown below). The addition of the third principal component only seems to highlight a potential outlier (sample 8, not shown). Potentially, this sample could be removed from the analysis, or, noted when doing further downstream analysis. The removal of outliers should be exercised with great caution and backed up with several other types of analyses (e.g. clustering) or graphical outputs (e.g. boxplots, heatmaps, etc).

```
# Interactive 3D plot will load the rgl library.
plotIndiv(final.pca.multi, style = '3d',
          group = multidrug$cell.line$Class,
          title = 'ABC transporters, PCA comp 1 - 3')
```

These plots suggest that the largest source of variation explained by the first two components can be attributed to the melanoma cell line, while the third component highlights a single outlier sample. Hence, the interpretation of the following outputs should primarily focus on the first two components.

Note:

- *Had we not scaled the data, the separation of the melanoma cell lines would have been more obvious with the addition of the third component, while PC1 and PC2 would have also highlighted the sample outliers 4 and 8. Thus, centering and scaling are important steps to take into account in PCA.*

9.5.3.5   Variable plot: Correlation circle plot

Correlation circle plots indicate the contribution of each variable to each component using the `plotVar()` function, as well as the correlation between variables (indicated by a 'cluster' of variables). Note that to interpret the latter, the variables need to be centered and scaled in PCA, see more details in Section 6.2.

```
plotVar(final.pca.multi, comp = c(1, 2),
        var.names = TRUE,
        cex = 3,            # To change the font size
        # cutoff = 0.5,     # For further cutoff
        title = 'Multidrug transporter, PCA comp 1 - 2')
```

The plot in Figure 9.5 highlights a group of ABC transporters that contribute to PC1: ABCE1, and to some extent the group clustered with ABCBs that contributes positively to PC1, while ABCAs contributes negatively. We also observe a group of transporters that contribute to both PC1 and PC2: the group clustered with ABCC2 contributes positively to PC2 and negatively to PC1, and a cluster of ABCC12 and ABCD2 that contributes negatively to both PC1 and PC2. We observe that several transporters are inside the small circle. However, examining the third component (argument `comp = c(1, 3)`) does not appear to reveal further transporters that contribute to this third component. The additional argument `cutoff = 0.5` could further simplify this plot.

9.5.3.6   Biplot: Samples and variables

A biplot allows us to display both samples and variables simultaneously to further understand their relationships (see Section 6.2). Samples are displayed as dots while variables are displayed at the tips of the arrows. Similar to correlation circle plots, data must be centered and scaled to interpret the correlation between variables (as a cosine angle between variable arrows).

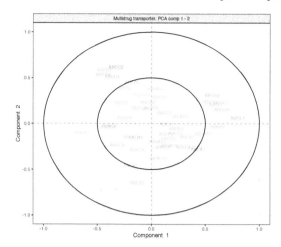

FIGURE 9.5: Correlation Circle plot from the PCA performed on the ABC.trans data. The plot shows groups of transporters that are highly correlated, and also contribute to PC1 – near the big circle on the right hand side of the plot (transporters grouped with those in orange), or PC1 and PC2 – top left and top bottom corner of the plot, transporters grouped with those in pink and yellow.

```
biplot(final.pca.multi, group = multidrug$cell.line$Class,
       legend.title = 'Cell line')
```

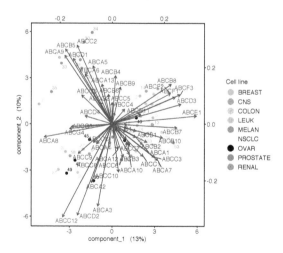

FIGURE 9.6: Biplot from the PCA performed on the ABS.trans data. The plot highlights which transporter expression levels may be related to specific cell lines, such as melanoma.

The biplot in Figure 9.6 shows that the melanoma cell lines seem to be characterised by a subset of transporters such as the cluster around ABCC2 as highlighted previously in Figure 9.5. Further examination of the data, such as boxplots (as shown in Figure 9.7), can further elucidate the transporter expression levels for these specific samples.

```
ABCC2.scale <- scale(X[, 'ABCC2'], center = TRUE, scale = TRUE)

boxplot(ABCC2.scale ~
        multidrug$cell.line$Class, col = color.mixo(1:9),
        xlab = 'Cell lines', ylab = 'Expression levels, scaled',
        par(cex.axis = 0.5), # Font size
        main = 'ABCC2 transporter')
```

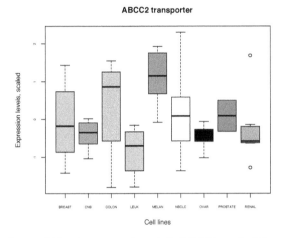

FIGURE 9.7: Boxplots of the transporter ABCC2 identified from the PCA correlation circle plot (Figure 9.5) and the biplot (Figure 9.6) show the level of ABCC2 expression related to cell line types. The expression level of ABCC2 was centered and scaled in the PCA, but similar patterns are also observed in the original data.

9.5.4 Example: Sparse PCA

In the `ABC.trans` data, there is only one missing value. Missing values can be handled by sPCA, as described in Section 5.3. However, if the number of missing values is large, we recommend imputing them with NIPALS, as described in the next section.

9.5.4.1 Choose the number of variables to select

First, we must decide on the number of components to evaluate. The previous tuning step indicated that `ncomp = 3` was sufficient to explain most of the variation in the data, which is the value we choose in this example. We then set up a grid of `keepX` values to test, which can be thin or coarse depending on the total number of variables. We set up the grid to be thin at the start, and coarse as the number of variables increases. The `ABC.trans` data includes a sufficient number of samples to perform repeated five-fold cross-validation to define the number of folds and repeats (refer to Section 7.2 for more details, leave-one-out CV is also possible if the number of samples N is small by specifying `folds = $N$`). The computation may take a while if you are not using parallelisation (see additional parameters in `tune.spca()`, here we use a small number of repeats for illustrative purposes. We then plot the output of the tuning function.

```
grid.keepX <- c(seq(5, 30, 5))
# grid.keepX  # To see the grid

set.seed(30) # For reproducibility with this handbook, remove otherwise
tune.spca.result <- tune.spca(X, ncomp = 3,
                              folds = 5,
                              test.keepX = grid.keepX, nrepeat = 10)

# Consider adding up to 50 repeats for more stable results
tune.spca.result$choice.keepX
```

```
## comp1 comp2 comp3
##    15    15    25
```

```
plot(tune.spca.result)
```

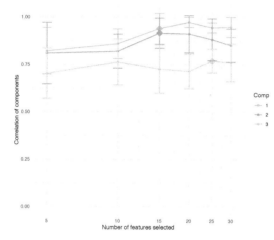

FIGURE 9.8: Tuning the number of variables to select with sPCA on the ABC.trans data. For a grid of number of variables to select indicated on the x-axis, the average correlation between predicted and actual components based on cross-validation is calculated and shown on the y-axis for each component. The optimal number of variables to select per component is assessed via one-sided t−tests and is indicated with a diamond.

The tuning function outputs the averaged correlation between predicted and actual components per `keepX` value for each component. It indicates the optimal number of variables to select for which the averaged correlation is maximised on each component (as described in Section 9.3). Figure 9.8 shows that this is achieved when selecting 15 transporters on the first component, and 15 on the second. Given the drop in values in the averaged correlations for the third component, we decide to only retain two components.

Note:

- *If the tuning results suggest a large number of variables to select that is close to the total number of variables, we can arbitrarily choose a much smaller selection size.*

9.5.4.2 Final sparse PCA

Based on the tuning above, we perform the final sPCA where the number of variables to select on each component is specified with the argument `keepX`. Arbitrary values can also be input if you would like to skip the tuning step for more exploratory analyses:

```
# By default center = TRUE, scale = TRUE
keepX.select <- tune.spca.result$choice.keepX[1:2]

final.spca.multi <- spca(X, ncomp = 2, keepX = keepX.select)

# Proportion of explained variance:
final.spca.multi$prop_expl_var$X
```

```
##      PC1     PC2
## 0.10107 0.08202
```

The amount of explained variance for sPCA is briefly described in Section 9.2. Overall when considering two components, we lose approximately 4.6 % of explained variance compared to a full PCA, but the aim of this analysis is to identify key transporters driving the variation in the data, as we show below.

9.5.4.3 Sample and variable plots

We first examine the sPCA sample plot:

```
plotIndiv(final.spca.multi,
          comp = c(1, 2),   # Specify components to plot
          ind.names = TRUE, # Show row names of samples
          group = multidrug$cell.line$Class,
          title = 'ABC transporters, sPCA comp 1 - 2',
          legend = TRUE, legend.title = 'Cell line')
```

In Figure 9.9, component 2 in sPCA shows clearer separation of the melanoma samples compared to the full PCA. Component 1 is similar to the full PCA. Overall, this sample plot shows that little information is lost compared to a full PCA.

A biplot can also be plotted that only shows the selected transporters (Figure 9.10):

```
biplot(final.spca.multi, group = multidrug$cell.line$Class,
       legend =FALSE)
```

The correlation circle plot highlights variables that contribute to component 1 and component 2 (Figure 9.11):

```
plotVar(final.spca.multi, comp = c(1, 2), var.names = TRUE,
        cex = 3, # To change the font size
        title = 'Multidrug transporter, sPCA comp 1 - 2')
```

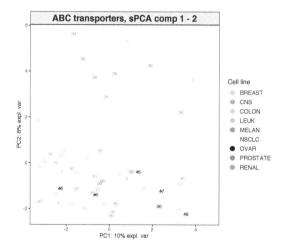

FIGURE 9.9: Sample plot from the sPCA performed on the ABC.trans data. Samples are projected onto the space spanned by the first two sparse principal components that are calculated based on a subset of selected variables. Samples are coloured by cell line type and numbers indicate the sample IDs.

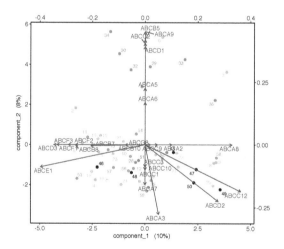

FIGURE 9.10: Biplot from the sPCA performed on the ABS.trans data after variable selection. The plot highlights in more detail which transporter expression levels may be related to specific cell lines, such as melanoma, compared to a classical PCA.

The transporters selected by sPCA are amongst the top important ones in PCA. Those coloured in green in Figure 9.5 (ABCA9, ABCB5, ABCC2 and ABCD1) show an example of variables that contribute positively to component 2, but with a larger weight than in PCA. Thus, they appear as a clearer cluster in the top part of the correlation circle plot compared to PCA. As shown in the biplot in Figure 9.10, they contribute in explaining the variation in the melanoma samples.

We can extract the variable names and their positive or negative contribution to a given component (here component 2), using the `selectVar()` function:

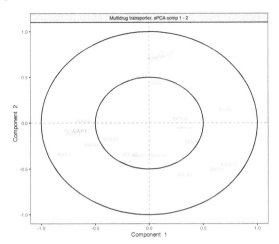

FIGURE 9.11: Correlation Circle plot from the sPCA performed on the ABC.trans data. Only the transporters selected by the sPCA are shown on this plot. Transporters coloured in green are discussed in the text.

```
# On the first component, just a head
head(selectVar(final.spca.multi, comp = 2)$value)
```

```
##          value.var
## ABCA9      0.4511
## ABCB5      0.4185
## ABCC2      0.4046
## ABCD1      0.3921
## ABCA3     -0.2780
## ABCD2     -0.2256
```

The loading weights can also be visualised with `plotLoading()`, as described in Section 6.2, where variables are ranked from the least important (top) to the most important (bottom) in Figure 9.12). Here on component 2:

```
plotLoadings(final.spca.multi, comp = 2)
```

9.5.5 Example: Missing values imputation

Missing values are coded as NA (which stands for 'Not Available') in R. In our context, missing values should be random, meaning that no entire row or column contains missing values (those should be discarded beforehand). Here we assume a particular value is missing due to technical reasons (and hence it is not replaced by a specified detection threshold value, see Section 1.4).

In Section 5.3, we detailed how the NIPALS algorithm can manage missing values. Here we illustrate the particular case where we *impute* (estimate) missing values using NIPALS on the `compound` data set. Missing value imputation is required for specific methods, including prediction, or, when methods such as NIPALS do not perform well when the number of

Loadings on comp 2

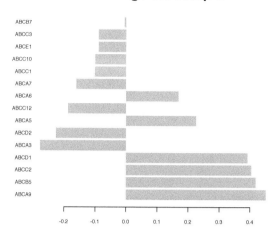

FIGURE 9.12: sPCA loading plot of the `ABS.trans` data for component 2. Only the transporters selected by sPCA on component 2 are shown, and are ranked from least important (top) to most important. Bar length indicates the loading weight in PC2.

missing values is rather large. The latter case can be quickly diagnosed as the proportion of explained variance does not decrease but increases with the number of components! Our experience has shown that starts to happen when the proportion of missing values is above 20%. We recommend to first remove the variables with too many missing values, and, if required in downstream analysis, to estimate the remaining missing values as illustrated in this Section on the `compound` data.

9.5.5.1 Data filtering

We first start by identifying which variables include a large number of missing values.

```
# Here, X.na is the compound data.
X.na <- multidrug$compound

# First, sum the number of missing values per variable.
sum.na.per.var <- apply(X.na, 2, function(x){sum(is.na(x))})

# A simple plot to show the NA rate per variable:
plot(sum.na.per.var/nrow(X), type = 'h',
     xlab = 'variable index', ylab = 'NA rate', main = 'NA rate per variable')
```

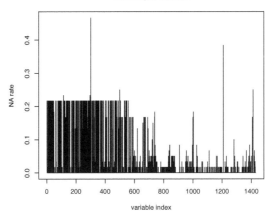

```
# These variables could be removed according to a threshold, e.g. 20%
remove.var <- which(sum.na.per.var/nrow(X) > 0.20)
X.na <- X.na[, -c(remove.var)]
dim(X.na) # Check the dimensions of the data
```

```
## [1]    60 1260
```

```
# Any additional information should also be updated to reflect the filtering, e.g.
compound.name.noNA <- multidrug$comp.name[-c(remove.var)]
```

We then calculate the proportion of missing values remaining, here a small amount:

```
sum(is.na(X.na)) / (length(X.na)) # Proportion of missing values
```

```
## [1] 0.01418
```

9.5.5.2   Imputation with NIPALS

Contrary to PCA that we use for dimension reduction, here when we use PCA-NIPALS we want to include as many components as possible to reconstruct the data as accurately as possible (you can go up to $\min(N, P)$ in the `mixOmics` implementation), and estimate missing values. Here we consider 10 components but more could be added.

```
# This might take some time - here 40s, depending on the size of the dataset.
# Reconstructed matrix from NIPALS:
X.nipals <- impute.nipals(X.na, ncomp = 10)
```

The reconstructed data matrix from NIPALS only replaces the missing values in the original dataset

Notes:

- *Sometimes a warning may appear for the calculation of a particular component if the algorithm struggles to converge when values are not missing by random, or if the components are not orthogonal to each other.*

- *By default this function will center the data for improved missing value imputation.*

9.5.5.3 Comparison of NIPALS vs PCA on imputed data

In this example, the number of missing values is small, allowing us to show that NIPALS without imputing missing values gives similar results to a classic PCA on data imputed with NIPALS. Thus, NIPALS can be efficient in managing missing values when data imputation is not required.

```
# Internally this function calls NIPALS but does not impute NAs
pca.with.na <- pca(X.na, ncomp = 3, center = TRUE, scale = TRUE)

# PCA on data imputed with NIPALS
pca.no.na <- pca(X.nipals, ncomp = 3, center = TRUE, scale = TRUE)
```

We can look at various outputs, such as the cumulative amount of explained variance:

```
pca.with.na$cum.var
```

```
##    PC1    PC2    PC3
## 0.2709 0.3539 0.4299
```

```
pca.no.na$cum.var
```

```
##    PC1    PC2    PC3
## 0.2684 0.3507 0.4251
```

The amount of explained variance is similar between both approaches.

We can also compare the sample plots (Figure 9.13):

```
plotIndiv(pca.with.na, group = multidrug$cell.line$Class, legend = TRUE,
          title = 'PCA with missing values')

plotIndiv(pca.no.na, group = multidrug$cell.line$Class, legend = TRUE,
          title = 'PCA with NIPALS imputed missing values')
```

The sample plots show similar trends. We can dig further by calculating the correlation between the principal components from both approaches:

```
cor(pca.with.na$variates$X, pca.no.na$variates$X)
```

```
##                PC1      PC2       PC3
## PC1 -0.9998291 -0.00341  0.001295
## PC2  0.0015388 -0.99806  0.052820
## PC3 -0.0003844 -0.05090 -0.997816
```

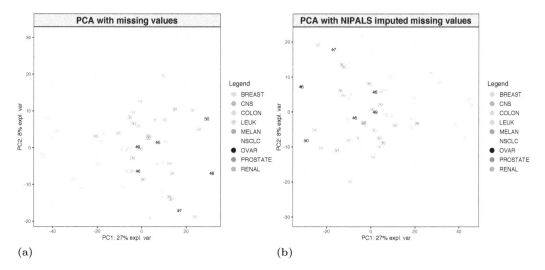

(a)   (b)

FIGURE 9.13: Sample plots comparison between PCA performed on the data with missing values, or PCA performed on NIPALS imputed data on the `compound` data. The plots show similar trends, even if the signs of the principal components are swapped.

A negative sign can occur as the principal component rotation can be inversed. The correlation between both sets of principal components is very high. Thus, these explorations indicate that the imputed data values are within the distribution of the original data, and do not increase the overall variance. Thus, if one needs full non-missing data, the imputed data set can be used for further analyses. The correlation circle plots are similar (not shown here).

9.6 To go further

This is only the start of your data exploration! Below we list potential further avenues to deepen the PCA exploratory analyses.

9.6.1 Additional processing steps

Several data processing steps can be considered depending on the data and the problem at hand, for example to manage repeated measurements or log ratio transformed compositional data (see Section 4.1). These additional data transformation steps have been included directly into the `pca()` and `spca()` functions using the arguments `multilevel` and `logratio`.

9.6.2 Independent component analysis

Whilst we consider PCA as a compulsory first step for the exploration of large data sets, other matrix factorisation techniques, such as Independent Component Analysis (ICA), can be considered, for example, when the *variance* is not the primary factor of interest in the data set. ICA is a blind source signal separation technique used to reduce the effects of noise or artefacts (Comon (1994), Hyvärinen and Oja (2001)). It assumes that a signal is generated from independent sources, and that variables with Gaussian distributions represent noise. ICA identifies non-Gaussian components that are statistically independent, i.e. there is no overlapping information between the components. ICA therefore involves high order statistics, while PCA involves second order statistics by constraining the components to be mutually orthogonal. As a result, PCA and ICA often project data into different subspaces. ICA faces limitations such as instability as it is based on a stochastic algorithm. Robust results can be obtained by running the method several times and averaging the results. The number of independent components to extract and choose is a hard outstanding problem, as independent components are not ordered according to relevance. Measures such as kurtosis, or ℓ_2 norm can be used. Despite these limitations, ICA can be considered as a successful alternative to PCA and has been increasingly used for mining biological data sets, and extracting gene modules (Sompairac et al. (2019), Saelens et al. (2018)).

We have developed an ICA variant, called Independent PCA (also for variable selection), which is detailed in Yao et al. (2012) and implemented in the functions `ipca()` and `sipca()`. Briefly, we use first PCA to reduce the dimensions of the data and generate the loading vectors. ICA is then applied on the PCA loading vectors as a denoising process before calculating the new Independent Principal Components. IPCA is appropriate when the loading vector distribution is non-Gaussian. In that case, we found that IPCA was able to summarise the information of the data better or with a smaller number of components than PCA or ICA.

9.6.3   Incorporating biological information

So far, we have only considered *data-driven* matrix factorisation. However, other variants of PCA can be used as pathway-based tools. For example, unlike PCA's components that are linearly defined, Principal Curves (Hastie and Stuetzle, 1989) fits smooth regression curves that are non-linear. Each curve can be considered as a local average of the data and is fitted based on the principal component to capture the variation in the data, and the projection of the samples on these curves can be interpreted as a new score. The reference samples (e.g. control) define the start of the principal curve direction. All samples are then projected onto this curve, and the distance between the projection and that of the centroid of the reference samples is defined as a new score for that sample. If we assume there is an effect of a specific biological condition, then this score can be considered as a dysregulation score for a particular sample, compared to the reference samples. This approach has been extended by including only genes belonging to specific and known pathways to construct these curves (Drier et al., 2013). One of our examples of such an analysis focusing on homologous recombination DNA repair pathways in sporadic breast tumours can be found in Liu et al. (2016).

Another approach is to consider other types of penalisation, for example, including a weighted group penalty on genes believed to belong to the same biological network (see Pan et al. (2010) and more recently Li et al. (2017)).

9.7 FAQ

Should I scale my data before running PCA? (`scale = TRUE` in `pca()`)

- Without scaling: a variable with high variance will be instrumental in defining the first principal component.
- With scaling: one noisy variable with low variance will be assigned similar variance as other (possibly) meaningful variables.

Can I perform PCA with missing values?

- Yes, PCA will run NIPALS in the background. Check that missing values comprise less than 20% of the total data, and that the PCA screeplot (amount of explained variance) decreases, otherwise, consider imputing missing values after some filtering, as explained in Section 9.5.

When do I need to estimate missing values?

In general we recommend to use this method only if:

- The downstream methods used to analyse the data cannot manage missing values and,
- Missing values comprise of less than 20% of the total data set.

When should I apply a multilevel approach in PCA? (`multilevel` argument in `pca()`)

- When the unique individuals are measured more than once (repeated measures).
- When the individual variation is larger than the treatment or time variation (this means that samples from each unique individual will tend to cluster rather than the treatments).
- When a multilevel vs. no multilevel seems to visually make a difference on a PCA plot.

More details can be found in Section 4.1 and Appendix 4.A.

When should I apply a CLR transformation in PCA? (`logratio = 'CLR'` argument in `pca()`)

When data are compositional, i.e. expressed as relative proportions. This is usually the case with microbiome studies, see more details in Section 4.1, Appendix 4.B and on our website[2].

What are the limitations of PCA?

Classical PCA can be limited when:

- The biological question is not related to the highest variance in the data (an alternative may be Independent Component Analysis or variants such as `ipca()` and `sipca()` in `mixOmics` (Yao et al., 2012), see Section 9.6 above.

- When too many noisy variables contribute to the variance (consider applying sPCA).

- When there are too many missing values (NIPALS will be used to run PCA, or to estimate missing values).

- When samples are **not** independent (e.g. time course data, repeated measures) and the subject variation is greater than the time/repeated treatment variation (consider using a multilevel decomposition in PCA).

---

[2]http://mixomics.org/mixmc/

9.8   Summary

PCA is a useful technique for displaying the major sources of variation in a data set and identifying whether such sources of variation correspond to biological conditions, batch effects or experimental bias. It performs dimension reduction, and in the case of sparse PCA, feature selection. When there are a small number of missing values in the data set, mixOmics employs the NIPALS algorithm rather than the standard, faster, SVD algorithm. When missing values need to be imputed, NIPALS can be used to estimate these missing values. Sample plots enable us to visualise how the samples are separating or grouping together, while correlation circle plots enable us to visualise variables that are the most influential in describing variation in the data, and how they relate to each other. Biplots enable us to further understand associations between samples and variables. We have illustrated these plots, along with other graphical and numerical outputs, to support the analyst in making sense of the data structure. Based on these preliminary explorations, more specific analyses can follow. Several other related PCA variants can also be considered to deepen these analyses.

9.A   Appendix: Non-linear Iterative Partial Least Squares

9.A.1   Solving PCA with NIPALS

This pseudo algorithm was presented in Section 5.3 but is included here for easier reference in this Appendix.

INITIALISE the first component t using SVD(X) or any column of X
  FOR EACH dimension
    UNTIL CONVERGENCE of a
     (a) Calculate loading vector $a = X^T t/t't$ and norm a to 1
     (b) Update component $t = Xa/a'a$
    DEFLATE $X = X - ta'$ as current matrix

This algorithm is called non-linear as it estimates parameters from a bilinear model with the local regressions. The division by $t't$ in step (a) allows us to define $X^T t$ as a regression slope – this is particularly useful when there are missing values. Similarly for step (b).

9.A.2   Estimating missing values with NIPALS

The local linear regressions for each PCA dimension in Steps (a) and (b) described in Section 5.3 enable us to understand how NIPALS can manage missing values: the absence of some values in the variable X^j should not affect the regression fit, if they are missing at random and not too numerous.

Missing values can also be estimated by reconstructing the data matrix $\boldsymbol{X}$ based on the loading vector $\boldsymbol{a}$ and component $\boldsymbol{t}$. For example, to estimate the missing values $\boldsymbol{X}_i^j$ for sample i and variable j, we calculate

$$\hat{\boldsymbol{X}}_i^j = \sum_{h=1}^{H} \boldsymbol{t}_i^h \boldsymbol{a}_j^h \qquad (9.3)$$

where $\hat{\boldsymbol{X}}_i^j$ is the estimation of the missing value, and $(\boldsymbol{t}^h, \boldsymbol{a}^h)$ is the component and loading vector pair in dimension h. We can see from this equation that the estimation is made to the order H that is the chosen dimension in NIPALS. This explains why NIPALS requires quite a large number of components in order to estimate these missing values.

Note:

- *In the case of missing values, the component $\boldsymbol{t}^h$ is initialised with any column of the current data matrix $\boldsymbol{X}$.*

9.B   Appendix: sparse PCA

9.B.1   sparse PCA-SVD

In `mixOmics` we have implemented the sPCA from Shen and Huang (2008) which is based on SVD that fits into a regression-type problem (as we highlighted in Section 5.3). Sparsity is achieved via ℓ_1 (lasso) penalisation estimated with soft-thresholding.

We define the Frobenius norm between $\boldsymbol{X}$ and its rank-l matrix $\boldsymbol{X}^{(l)}$ as[3]:

$$||\boldsymbol{X} - \boldsymbol{X}^{(l)}||_F^2 = \text{trace}\{(\boldsymbol{X} - \boldsymbol{X}^{(l)})(\boldsymbol{X} - \boldsymbol{X}^{(l)})^T\},$$

The closest rank-l matrix approximation to $\boldsymbol{X}$ in terms of the squared Frobenius norm is:

$$\boldsymbol{X}^{(l)} \equiv \sum_{k=1}^{l} d_k \boldsymbol{u}^k \boldsymbol{a}^{k'}. \qquad (9.4)$$

We have seen this equation already in Equation (9.3) to estimate missing values. Here we obtain the best rank-1 approximation of $\boldsymbol{X}$ by seeking the $N-$ and $P-$ dimensional vectors $\boldsymbol{u}$ and $\boldsymbol{a}$ (both constrained to be of norm 1):

$$\min_{||\boldsymbol{u}||=||\boldsymbol{a}||=1} ||\boldsymbol{X} - \boldsymbol{u}\boldsymbol{a}'||_F^2,$$

which we can solve with the first left and right singular vectors $(\boldsymbol{t}^1, \boldsymbol{a}^1)$ from the SVD where $\boldsymbol{t}^1 = d_1 \boldsymbol{u}^1$ and $\boldsymbol{a} = \boldsymbol{a}^1$. The second set of vectors will give the best approximation of the rank-2 of $\boldsymbol{X}$ (or equivalently of the matrix $\boldsymbol{X} - d_1 \boldsymbol{u}^1 \boldsymbol{a}^{1'}$).

Steps (a) and (b) in the NIPALS algorithm 9.A detail how least squares regressions are performed in PCA NIPALS to obtain $\boldsymbol{a}$ as a regression of $\boldsymbol{X}$ on a fixed $\boldsymbol{t}$. In such a regression context, it is then possible to apply regularisation penalties on $\boldsymbol{a}$ to obtain a *sparse* loading vector to perform variable selection. The objective function of sPCA can be written as:

---

[3]The Frobenius norm is an extension of the Euclidean norm for matrices.

$$\underset{t,a}{\mathrm{argmin}} \ ||\boldsymbol{X} - \boldsymbol{ta'}||_F^2 + P_\lambda(\boldsymbol{a}), \tag{9.5}$$

where $P_\lambda(\boldsymbol{a}) = \sum_{j=1}^P P_\lambda(|\boldsymbol{a}_j|)$ is a penalisation term with parameter λ that is applied on each element of the vector $\boldsymbol{a}$. The sPCA version in `mixOmics` implements the soft-thresholding penalty as an estimate of the lasso[4] (Friedman et al., 2007). The penalty is applied on each variable i as follows:

$$P_\lambda(a_j) = \mathrm{sign}(a_j)(|a_j| - \lambda)_+ \tag{9.6}$$

with $(x)_+ \Leftrightarrow x = 0$ if $x \leq 0$ and $x = x$ otherwise. The penalisation results in a loading vector with many 0 values for the variables that are considered irrelevant for the local regression in step (a). Since $\boldsymbol{t} = X\boldsymbol{a}$, the principal component $\boldsymbol{t}$ is now calculated on a subset of relevant variables with non-zero weights in the sparse loading vector. The selected variables can be extracted by looking at the non-zero elements in the loading vector (see the function `select.var()`).

In `mixOmics` the penalty λ has been replaced by the number of variables to select (`keepX`) for practical reasons, but both criteria are equivalent (see Section 9.3 for tuning this parameter).

9.B.2   sPCA pseudo algorithm

The pseudo code below explains how to obtain a sparse loading vector associated to the first principal component.

INITIALISE $\boldsymbol{t} = d\boldsymbol{u}$ and $\boldsymbol{a} = \boldsymbol{a}$ from the first singular value and vectors of SVD($\boldsymbol{X}$) or any column of $\boldsymbol{X}$
   FOR EACH dimension
     UNTIL CONVERGENCE of $\boldsymbol{a}$
     (a) Calculate sparse loading vector $\boldsymbol{a} = P_\lambda(\boldsymbol{X}^T\boldsymbol{t})$ and norm $\boldsymbol{a}$ to 1
     (b) Update component $\boldsymbol{t} = X\boldsymbol{a}$
     DEFLATE $\boldsymbol{X} = \boldsymbol{X} - \boldsymbol{ta'}$ as the current matrix

The sparse loading vectors are computed component-wise which results in a list of selected variables per component. The deflation step should ensure that the loading vectors are orthogonal to each other (i.e. different variables selected on different dimensions), but that also depends on the degree of penalty that is applied.

9.B.3   Other sPCA methods

Several sPCA methods have been proposed and implemented in the literature. Each of these methods differ depending on the formulation of sPCA and the type of penalty applied. For example a convex sparse PCA method that reformulates the PCA problem as a regression problem and imposes an elastic net penalty on the loading vectors has been proposed (Zou et al., 2006), or alternatively, a penalised matrix decomposition with a lasso penalty on the loadings (Witten et al., 2009). In `mixOmics`, we used the formulation from Shen and Huang

---

[4]This is equivalent to a pathwise coordinate descent optimization algorithm.

(2008) with a Lagrange form to constrain the loading vectors. In practice, both approaches from Witten et al. (2009) and Shen and Huang (2008) (and thus ours) lead to very similar results, whilst, as expected, almost half the selected variables differed when comparing these approaches and the elastic net approach from Zou et al. (2006) when analysing the same data set.

Among them, we can list the functions:

- `spca()` and `arrayspc()` from the `elasticnet` package (Zou and Hastie, 2018). `spca()` takes as input either a data matrix or a covariance / correlation matrix and includes either a penalty parameter or the number of variables to select on each dimension, whereas `arrayspc()` uses iterative SVD and soft-thresholding on large data sets. Note that none of these functions output the components, and no visualisation is available.

- `SPC()` from the `PMA` package (Witten and Tibshirani, 2020). The function takes as an argument the penalty parameter. The sparse singular vectors are output and a tuning function is also available (as detailed below). An interesting feature is the ability for the components to be orthogonal by re-projecting the sparse principal component in an orthogonal space. The package does not offer visualisations.

9.B.3.1 Tuning sPCA

In `mixOmics`, our proposed criterion is based on cross-validation subsampling and calculating the correlation between predicted and actual components on test samples, as summarised in Section 9.3.

As in all our tuning functions, we advise to choose at least 50 repeats across the CV runs.

INITIALISE `grid.keepX = NULL`
FOR `comp = 1`
  FOR EACH `keepX` value from a specified grid of values
    FOR EACH `repeat` and `fold`, define the X_{train} and X_{test} sets
        (a) Run `spca_train <- spca(`X_{train}`, keepX = c(grid.keepX, keepX), ncomp = comp)` and extract a_{train}
        (b) Calculate the predicted component $t_{\text{test}} = X_{\text{test}} a_{\text{train}}$
        (c) Run `spca_full <- spca(`X`, keepX = c(grid.keepX, keepX), ncomp = comp)` and extract t
        (d) Caculate $\rho = \text{cor}(t_{\text{test}}, t[\text{test samples}])$ for each `fold`, `repeat` and `keepX` values.
     – Average ρ_{keepX} across folds and repeats, set `keepX.optim` as the `keepX` value that maximises ρ_{keepX}
     – Update `grid.keepX = c(grid.keepX, keepX.optim)` to move to the next dimension.

To go beyond `comp > 1`, we then set `comp = comp + 1` and run the algorithm. In Step (b), we first calculate $a_{\text{test}} = X^T_{\text{test}} t_{\text{test}}$, then deflate $X_{\text{test}} = X_{\text{test}} - t_{\text{test}} a'_{\text{test}}$.

9.B.3.2 Tuning sPCA in other packages

Several tuning criteria have been proposed in the different R packages available.

The tuning function from Witten et al. (2009) in the `PMA` package (Witten and Tibshirani, 2020) is based on a rank-1 approximation matrix and thus only considers the tuning for the first dimension using repeated CV (deflation of the current matrix is then required to go to the next dimension). A fraction of the data is replaced with missing values and sPCA is performed on the new data matrix using a range of tuning parameter values (the λ penalisation parameter). The mean squared error (MSE) of the rank-1 imputation of the missing values compared to the full data is then calculated. The optimal penalty parameter is chosen so that it minimises the average MSE across folds and repeats. The proposition of creating artificial missing values and imputing them using rank-1 approximation matrix is appealing. The potential limitations we have identified include the choice of the fraction of missing values (currently set by default but the value is not specified), and the tendency of the MSE to select a large number of variables, which goes against a parsimonious model.

Zou et al. (2006) proposed to choose the penalisation parameter based on the percentage of explained variance (PEV). In their example (Section 5.3 in Zou et al. (2006)), the authors show that the PEV decreases at a slow rate when the sparsity increases, reporting a loss of 6% PEV compared to a non sparse PCA when selecting 2.5% of the original 16,063 genes. However, when testing on the `multidrug ABC.trans` or the `srbct` expression data, we observed a loss from 30% to 47% of the PEV, with a sharp PEV decrease when sparsity increased.

Shen and Huang (2008) proposed two tuning criteria. The first approach is based on the minimisation of the squared Frobenius norm of rank-1 matrix approximation using CV (in that case the predicted components are calculated based on the loading vector from the training set, an approach similar to our proposed method). The second ad-hoc approach is similar to Zou et al. (2006), with the exception that the explained variance is adjusted to account for non-orthogonality between sparse principal components (note that we have not encountered such a case when the data are scaled in `spca()`).

Sill et al. (2015) proposed to use the stability criterion from Meinshausen and Bühlmann (2010) in sPCA. Using subsampling, selection probabilities for each variable can be estimated as the proportion of subsamples where a variable is selected in sPCA for a given range of possible penalisation parameters λ. Variables are then ranked according to their selection probability and a forward selection procedure is then applied, starting with the variable with the highest selection probability. Variables are then subsequently added to calculate sPCA. The forward selection procedure requires the use of a generalised information criterion (Kim et al., 2012). The method is likely to be computationally intensive, and is available as an R package `s4vdpca` on GitHub[5].

---

[5]https://github.com/mwsill/s4vdpca

10

Projection to Latent Structure (PLS)

10.1 Why use PLS?

Projection to Latent Structures (PLS), also called Partial Least Squares regression (Wold (1966), Wold et al. (2001)) is a multivariate methodology which relates, or *integrates*, two data matrices X (e.g. transcriptomics) and Y (e.g. metabolites). PLS can describe common patterns between two omics by identifying underlying factors in both data sets that best describe this pattern. Unlike traditional multiple regression models, it is not limited to uncorrelated variables: it can manage many noisy, collinear (correlated) and missing variables and can also simultaneously model several response variables Y. PLS is also a flexible algorithm that can address different types of integration problems: for this reason it is the backbone of most methods in `mixOmics`. This chapter presents several variants of PLS, both supervised (regression) and unsupervised, their numerical and graphical outputs, and their application in several examples.

10.1.1 Biological questions

My two omics data sets are measured on the same samples, where each data set (denoted X and Y) contains variables of the same type. Does the information from both data sets agree and reflect any biological condition of interest? If I consider Y as phenotype data, can I predict Y given the predictor variables X? What are the subsets of variables that are highly correlated and explain the major sources of variation across the data sets?

10.1.2 Statistical point of view

We wish to integrate two types of continuous data with either an exploratory or prediction-based question (introduced in Section 2.4). PLS aims to reduce the dimensions of each of the data sets via latent components by revealing the maximum covariance or correlation between these components (Section 3.1). PLS takes information from *both data sets* into account when decomposing each matrix into these latent components. Thus, we identify the latent components that best explain X *and* Y, while having the strongest possible relationship. Specifically, PLS can:

- Predict the response variables Y from the predictor variables X, where prior biological knowledge indicates which type of omics data is expected to explain the other type (PLS *regression mode*).

- Model variables from both data sets that covary (i.e. 'change together') across different conditions, where there is no biological *a priori* indication that one type of data may cause differences in the other (PLS *canonical mode*).

- Identify a subset of variables that best describe these latent components through penalisations to improve interpretability (*sparse* PLS, see Section 3.3).

10.2   Principle

PLS is a multivariate projection-based method: unlike PCA which maximises the variance in a single data set, PLS maximises the covariance between two data sets while seeking for linear combinations of the variables from both sets. These linear combinations are called *latent variables* or *latent components*. The weight vectors that are used to compute the linear combinations are called the *loading vectors*. Both latent variables and loading vectors come in pairs (one for each data set), as shown in Figure 10.1. The *dimensions* are the number of pairs of latent variables (each associated to a data set) necessary to summarise the covariance in the data.

In this chapter, we will use the following notations: $\boldsymbol{X}$ is an $(N \times P)$ data matrix, and $\boldsymbol{Y}$ is an $(N \times Q)$ data matrix, where N is the number of samples (or individuals), and P and Q are the number of variables (parameters) in X and Y. We denote by $\boldsymbol{X}^j$ and $\boldsymbol{Y}^k$ the vector variables in the $\boldsymbol{X}$ and the $\boldsymbol{Y}$ data sets ($j = 1, \ldots, P$ and $k = 1, \ldots, Q$).

PLS finds the latent components by performing successive regressions, as explained in Section 5.1, using projections onto the latent components to highlight underlying biological effects. PLS goes beyond a simple regression problem since $\boldsymbol{X}$ and $\boldsymbol{Y}$ are simultaneously modelled by successive local regressions, thus avoiding computational issues that arise when inverting large singular covariance matrices. PLS has three simultaneous objectives:

- To identify the best explanation of the $\boldsymbol{X}$-space through dimension reduction.
- To identify the best explanation of the $\boldsymbol{Y}$-space through dimension reduction.
- To identify the maximal relationship between the $\boldsymbol{X}$- and $\boldsymbol{Y}$-space through their covariance.

The objective function maximises the covariance between each linear combination of the variables from both data sets. For the first dimension, we solve:

$$\operatorname*{argmax}_{||\boldsymbol{a}||=1,\ ||\boldsymbol{b}||=1} \operatorname{cov}(\boldsymbol{X}\boldsymbol{a}, \boldsymbol{Y}\boldsymbol{b}). \tag{10.1}$$

The loading vectors are the pair of vectors $(\boldsymbol{a}, \boldsymbol{b})$ and their associated latent variables $(\boldsymbol{t}, \boldsymbol{u})$, respectively, where $\boldsymbol{t} = \boldsymbol{X}\boldsymbol{a}$ and $\boldsymbol{u} = \boldsymbol{Y}\boldsymbol{b}$. Similar to PCA, the loading vectors are directly interpretable, and indicate how the $\boldsymbol{X}^j$ and $\boldsymbol{Y}^k$ variables can explain the covariance between $\boldsymbol{X}$ and $\boldsymbol{Y}$. The latent variables $(\boldsymbol{t}, \boldsymbol{u})$ contain the information regarding the similarities or dissimilarities between individuals or samples.

The objective function for any dimensions h is:

$$\operatorname*{argmax}_{||\boldsymbol{a}^h||=||\boldsymbol{b}^h||=1, \boldsymbol{a}^{h\prime}\boldsymbol{a}^k=\boldsymbol{b}^{h\prime}\boldsymbol{b}^k=0, h<k} \operatorname{cov}(\boldsymbol{X}^{(h-1)}\boldsymbol{a}^h, \boldsymbol{Y}^{(h-1)}\boldsymbol{b}^h) \tag{10.2}$$

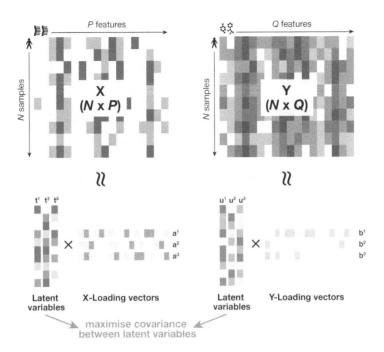

FIGURE 10.1: Schematic view of the PLS matrix decomposition of X and Y into sets of latent variables and loading vectors. Latent variables, or components, are built based on the original (centered and scaled) X and Y variables that are linearly combined according to the coefficient weights in their corresponding loading vectors. The loading vectors are defined so that the covariance between components is maximised, thus enabling data integration. The components are orthogonal within each data set. The PLS dimension is the number of pairs of components or loading vectors sufficient to summarise most of the covariance between data sets.

where $X^{(h-1)}$ and $Y^{(h-1)}$ are the residual matrices for dimension $h-1$ ($h = 1, \ldots, r$, where r is the rank of the matrix $X^T Y$). The residual matrices are calculated during the deflation step, as described in the PLS algorithm in Appendix 10.A. Each loading vector a^h is orthogonal to any other previous loading vector, and similarly for b^h.

10.2.1 Univariate PLS1 and multivariate PLS2

Numerous PLS variations have been developed depending on the 'shape' of the data Y (Figure 10.2) as well as the modelling purpose. Therefore, it is necessary to first define the type of analysis that is required. We will mainly refer to:

- **PLS1** for univariate analysis, where the response y is a single variable, and,
- **PLS2** for multivariate analysis, where the response Y is a matrix including more than one variable.

For both cases, X is a multi-predictor matrix. We will also consider two types of modelling in PLS2 that refers to different types of deflation (Figure 10.3):

- The **PLS regression mode** models a *uni-directional*, or *asymmetrical* relationship between two data sets,

- The **PLS canonical mode** models a *bi-directional* or *symmetrical* relationship between two data sets.

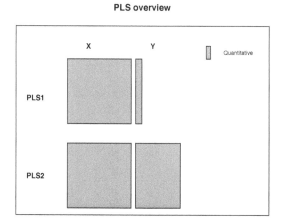

FIGURE 10.2: The difference between PLS1 and PLS2 depends on whether Y includes one or more variables. Here we consider quantitative X and Y data sets.

These PLS variants can be solved using either SVD of the matrix product $X^T Y$, or the iterative PLS algorithm. In `mixOmics` we use the SVD version proposed by Wegelin et al. (2000). These algorithms are presented in Appendix 10.A.

10.2.2 PLS deflation modes

The PLS modes (regression or canonical) aim to address different types of analytical questions. The way the matrices X and Y are deflated defines the PLS mode (as we introduced in Section 5.3).

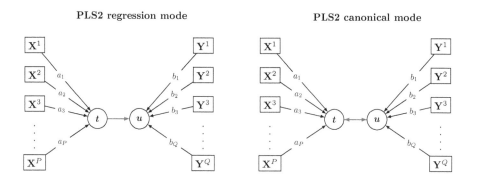

FIGURE 10.3: Difference between the two PLS2 modes, regression or canonical, using a PLS-path modelling graphical representation. The regression mode fits an asymmetric relationship between X and Y, where X is the predictor matrix and Y the response matrix, whereas the canonical mode fits a symmetric relationship between X and Y. The modes start to differ in the matrices deflation from the second dimension. Consequently, the latent variables (t, u) and loading vectors (a, b) are identical only on the first dimension for both modes.

10.2.2.1 Regression mode

PLS regression mode fits a linear relationship between multiple responses in the $\boldsymbol{Y}$ data set with multiple predictors from the $\boldsymbol{X}$ data set (multiple multivariable regression). This is the most well-known use for PLS. Examples include predicting a response to drug therapy given gene expression data (Prasasya et al., 2012), or inferring transcription factor activity from mRNA expression and DNA-protein binding measurements (Boulesteix and Strimmer, 2005). In this mode, the $\boldsymbol{Y}$ matrix is deflated with respect to the information that is extracted and modelled from the local regression on $\boldsymbol{X}$ (while $\boldsymbol{X}$ is deflated with respect to the information from $\boldsymbol{X}$). In this mode $\boldsymbol{Y}$ and $\boldsymbol{X}$ play an *asymmetric* role. Consequently, the latent variables to model $\boldsymbol{Y}$ from $\boldsymbol{X}$ would be different from those to model $\boldsymbol{X}$ from $\boldsymbol{Y}$. In general, the number of variables $\boldsymbol{Y}^k$ to predict are fewer in number than the predictors $\boldsymbol{X}^j$. We will illustrate a PLS2 regression mode analysis in Section 10.5.

10.2.2.2 Canonical mode

PLS canonical mode models a *symmetrical* linear relationship between $\boldsymbol{X}$ and $\boldsymbol{Y}$. The approach is in essence similar to Canonical Correlation Analysis (that will be covered in Chapter 11) but for a high dimensional setting. Although not widely used yet, the method has been used to highlight complementary information from transcripts measured on different types of microarrays (cDNA and Affymetrix arrays) for the same cell lines (Lê Cao et al., 2009). It is applicable when there is no *a priori* relationship between the two data sets, or in place of Canonical Correlation Analysis for large data sets, or when variable selection is required. In this mode, the $\boldsymbol{Y}$ matrix is deflated with respect to the information extracted and modelled from the local regression on $\boldsymbol{Y}$, and $\boldsymbol{X}$ is deflated with respect to the information from $\boldsymbol{X}$.

10.2.2.3 How do the different PLS modes differ?

For the first PLS dimension $h = 1$, all PLS modes output the same latent variables and loading vectors. As soon as the dimension h increases, these vectors will differ, as the matrices $\boldsymbol{X}$ and $\boldsymbol{Y}$ are deflated differently (and the biological questions asked are different).

10.2.2.4 Iterative local regressions and deflations to solve PLS

The PLS algorithm is detailed in Appendix 10.A. We summarise here the main ideas, which extend the concepts we have already covered from the PCA NIPALS in Section 5.3:

- The PLS algorithm is performed iteratively, one dimension h at a time, to obtain a pair of components $(\boldsymbol{t}, \boldsymbol{u})$ to which correspond the loading vectors $\boldsymbol{a}$ and $\boldsymbol{b}$ respectively, for each dimension h (Figure 10.1),

- The loading vector $\boldsymbol{a}$ $(\boldsymbol{b})$ is obtained by regressing the $\boldsymbol{X}$ $(\boldsymbol{Y})$ variables on the component $\boldsymbol{t}$ $(\boldsymbol{u})$. The components are then updated as $\boldsymbol{t} = \boldsymbol{X}\boldsymbol{a}$ and $\boldsymbol{u} = \boldsymbol{Y}\boldsymbol{b}$[1] (Figure 10.18),

- The deflation step enables the components and loading vectors to be orthogonal and defines the PLS modes. In a regression mode, the residual (current) matrix $\boldsymbol{Y}$ is calculated by subtracting the information related to $\boldsymbol{t}$ (associated to the $\boldsymbol{X}$ data set), whereas in a

---

[1] This can also be seen as a local regression of the $\boldsymbol{X}$ and $\boldsymbol{Y}$ samples on $\boldsymbol{a}$ and $\boldsymbol{b}$, respectively.

canonical mode we subtract the information related to $\boldsymbol{u}$. For both cases, the residual matrix $\boldsymbol{X}$ is obtained by subtracting the information related to $\boldsymbol{t}$.

Note:

- *PLS can manage missing values because local regressions are fitted on the available data, similar to NIPALS (see Section 5.3).*

10.2.3 sparse PLS

Even though PLS is highly efficient in a high dimensional context, the interpretability of PLS needed improvement. sPLS was developed to perform simultaneous variable selection in both data sets $\boldsymbol{X}$ and $\boldsymbol{Y}$, by including lasso penalisations in PLS on each pair of loading vectors $(\boldsymbol{a}, \boldsymbol{b})$ on each dimension (Lê Cao et al., 2008). The key is to solve PLS in a SVD framework to fit into a regression-type of problem to penalise the loading vectors. The algorithm is solved iteratively with the PLS algorithm, as detailed in Appendix 10.B.

Similarly to sparse PCA (Section 9.2), irrelevant variables from each loading vector are assigned a zero weight, and latent components are calculated only on a subset of 'relevant' variables. The zero weights are assigned optimally with lasso, whilst ensuring that the covariance between data sets is maximised. Both regression and canonical modes are available.

10.3 Input arguments and tuning

By default in PLS, the variables are centered *and* scaled as we are comparing two data sets measured on different platforms and scales (refer to Section 8.1 for more details). The parameters are similar to those we have seen in PCA and sPCA, namely, the number of dimensions `ncomp` and the number of variables to select (here `keepX` and `keepY`) per dimension and data set. Different tuning criteria are proposed depending on the type of parameter and the PLS mode.

10.3.1 The deflation mode

Two broad types of approaches can be used to choose the mode of PLS:

- Knowledge driven according to the biological assumptions: for example, it may make sense to consider proteomics data as the response variables $\boldsymbol{Y}$ and gene expression data as the predictor variables $\boldsymbol{X}$ in PLS regression mode while proteomics and lipids data might be more suitable in PLS canonical mode.

- Data driven according to numerical criteria that inform us of the quality of the fit of PLS: for example, we can compare the Q^2 criterion described for both modes and decide on the best fit. Note however that the numerical criteria we propose can be limited and inconclusive. As we are using a hypothesis-free approach, an alternative exploratory analysis with a canonical mode can be adopted.

10.3.2 The number of dimensions

The Q^2 measure has been proposed based on a cross-validation approach and extended for the PLS regression mode (Wold, 1982; Tenenhaus, 1998). The principle is to assess whether the addition of a dimension is beneficial in predicting the values of the data.

For the regression mode, we use cross-validation explained in Section 7.2 to *predict* $\hat{\boldsymbol{Y}}_{\text{test}}$ on the test set for dimension h, which we compare to the actual *fitted* values (from the $\boldsymbol{Y}$ data set) in the previous dimension $h - 1$. We detail the concepts of fitted vs. predicted values in Appendix 10.C.

For the canonical mode, we extend the same principles, except that Q^2 is calculated based on fitted and predicted values from the $\boldsymbol{X}$ data set, as described in Appendix 10.C.

The Q^2 criterion is a global measure that applies to both PLS1 and PLS2 and is calculated per dimension h (denoted Q^2_h). We can decide to retain a dimension h if $Q^2_h \geq 0.975$, a (somewhat) arbitrary threshold used in the SIMCA-P software (Umetri, 1996). A negative value of Q^2_h indicates a poor fit of the PLS model on the data - see further details in Appendix 10.C.

Notes:

- *The Q^2_1 value assesses the addition of dimension 2 compared to dimension 1 and will differ between the two PLS modes from the start.*

- *If the Q^2 criterion indicates a poor fit for either mode, we recommend choosing an arbitrary but small number of components and taking an exploratory approach that fits your biological question.*

10.3.3 Number of variables to select

10.3.3.1 PLS1

For PLS1, we can assess for the number of variables that gives the best prediction accuracy. We can use common measures of accuracy available in multivariate regression using cross-validation in the **tune()** function, namely: Mean Absolute Error (MAE) and Mean Square Error (MSE) that both average the model prediction error, Bias, and R^2, as described in Section 7.3. The **tune()** function for PLS1 outputs the optimal number of variables **keepX** to retain in the $\boldsymbol{X}$ data set.

Note:

- *MAE measures the average magnitude of the errors without considering their direction. It averages the absolute differences between the $\hat{\boldsymbol{Y}}$ predictions and the actual $\boldsymbol{Y}$ values over the fold test samples. The MSE also measures the average magnitude of the error, but the errors are squared before they are averaged. Thus, MSE tends to give a relatively high weight to large errors. Bias is the average of the differences between the $\hat{\boldsymbol{Y}}$ predictions and the actual $\boldsymbol{Y}$ values, and the R^2 is the average of the squared correlation coefficients between the predictions and the observations.*

10.3.3.2 PLS2

The measures of accuracy presented for PLS1 do not generalise well for PLS2, as we want to select the optimal number of variables with the best prediction accuracy from *both* X and Y simultaneously for each dimension. In `mixOmics`, we adopt a similar approach to sPCA described in Section 9.3 and Appendix 9.B.

Using repeated cross-validation from the `tune.spls()` function, and for a specified combination of (`keepX, keepY`) values, we calculate the predicted components t_{test} and u_{test}, which we compare to their respective fitted values t and u, either based on their correlation (`measure.tune = 'cor'`), which we want to maximise, or their squared differences (`measure.tune = 'RSS'`), which we want to minimise. The Residual Sum of Squares (RSS) criterion tends to select a smaller number of variables than the correlation criterion.

For a regression mode, we choose the best (`keepX, keepY`) pair based on the absolute correlation between u_{test} and u (since u best summarises Y), or their RSS.

For a canonical mode, we choose the best `keepX` based on (t_{test}, t), since t best summarises X, and the best `keepY` based on (u_{test}, u). More details are given in Appendix 10.A.

Notes:

- *These criteria were developed to guide the analysis, but do not represent absolute truth, and should be used with caution. Their optimal result may not correspond to the most biologically relevant result, in which case it may be more appropriate to choose your own parameters that are guided by biological knowledge.*

- *Keep practicality in mind when choosing the grid of* `keepX` *and* `keepY` *values: the number of chosen variables should match your biological question (smaller for guided analyses or larger for exploratory approaches but overall manageable for further interpretation).*

10.4 Key outputs

10.4.1 Graphical outputs

A variety of graphical outputs are available and further illustrated below. In particular, and as described in Chapter 6, the projection of the samples into the space spanned by the $X-$ or $Y-$ components enables us to visualise the agreement between both data sets, extracted by PLS. The correlation circle plot gives some insight into the correlation between variables from both data sets, which can also be emphasised further with relevance networks and clustered image maps (see Section 6.2 and Appendix 6.A). Finally, the simultaneous interpretation of both sample and variable plots enables us to understand the key variables that best characterise a particular set of samples, which can be visualised with a biplot.

10.4.2 Numerical outputs

The accuracy measures used to tune the PLS and sPLS parameters (including error measures and Q^2), can be output based on the final model using the `perf()` function. Other criteria can also be insightful:

- The stability of the variables selected can be obtained during the cross-validation process of the `perf()` function, as we described previously in Section 7.4. The frequency of selection occurrence is output separately for each type of variable.

- The amount of explained variance is based on the concept of *redundancy* defined in Tenenhaus (1998) and is calculated based on the components associated with each data set. For example, the variance explained by component t^h in X for a given dimension h is defined as:

$$Rd(\boldsymbol{X}, \boldsymbol{t}^h) = \frac{1}{P} \sum_{j=1}^{P} \mathrm{cor}^2(\boldsymbol{X}^j, \boldsymbol{t}^h)$$

where $\boldsymbol{X}^j$ is the variable j in $\boldsymbol{X}$ that is centered and scaled. Similarly for $\boldsymbol{Y}$, we calculate $Rd(\boldsymbol{Y}, \boldsymbol{u}^h)$.

- The Variable Importance in the Projection (VIP) assesses the contribution of each variable $\boldsymbol{X}^j$ in explaining $\boldsymbol{Y}$ through the components $\boldsymbol{t}$. It is defined for dimension h and variable $\boldsymbol{X}^j$ as:

$$\mathrm{VIP}_{hj} = \sqrt{P * \frac{\sum_{l=1}^{h} Rd(\boldsymbol{Y}, \boldsymbol{t}^l)(\boldsymbol{a}_j^l)^2}{Rd(\boldsymbol{Y}, \boldsymbol{t}^l)}}$$

using the definitions of Rd as above. Thus, for a given dimension h, the VIP takes into account the squared loading weight of a variable $\boldsymbol{X}^j$, $(\boldsymbol{a}_j^h)^2$, multiplied by the amount of variance explained in $\boldsymbol{Y}$ by the components $\boldsymbol{t}^h$. The ratio of the full amount of explained variance up to component h is calculated. This criterion is used in SIMCA-P software (Umetri, 1996) and considers a VIP > 1 as important in explaining $\boldsymbol{Y}$.

10.5 Case study: Liver toxicity

The data come from a liver toxicity study in which 64 male rats were exposed to non-toxic (50 or 150 mg/kg), moderately toxic (1500 mg/kg) or severely toxic (2000 mg/kg) doses of acetaminophen (paracetamol) (Bushel et al., 2007). Necropsy was performed at 6, 18, 24, and 48 hours after exposure and the mRNA was extracted from the liver. Ten clinical measurements of markers for liver injury are available for each subject. The microarray data contain expression levels of 3,116 genes. The data were normalised and preprocessed by Bushel et al. (2007).

`liver toxicity` contains the following:

- `$gene`: A data frame with 64 rows (rats) and 3,116 columns (gene expression levels),
- `$clinic`: A data frame with 64 rows (same rats) and 10 columns (10 clinical variables),
- `$treatment`: A data frame with 64 rows and 4 columns, describing the different treatments, such as doses of acetaminophen and times of necropsy.

We can analyse these two data sets (genes and clinical measurements) using sPLS1, then sPLS2 with a regression mode to explain or predict the clinical variables with respect to the gene expression levels.

10.5.1   Load the data

```
library(mixOmics)
data(liver.toxicity)
X <- liver.toxicity$gene
Y <- liver.toxicity$clinic
```

As discussed in Section 8.4, always ensure that the samples in the two data sets are in the same order, or matching, as we are performing data integration:

```
head(data.frame(rownames(X), rownames(Y)))
```

```
##    rownames.X. rownames.Y.
## 1        ID202       ID202
## 2        ID203       ID203
## 3        ID204       ID204
## 4        ID206       ID206
## 5        ID208       ID208
## 6        ID209       ID209
```

10.5.2   Quick start

Here are the key functions for a quick start, using the functions pls(), spls() and the graphic visualisations plotIndiv() and plotVar().

10.5.2.1   PLS

```
pls.result <- pls(X, Y)     # 1 Run the method
plotIndiv(pls.result)       # 2 Plot the samples
plotVar(pls.result)         # 3 Plot the variables
```

Let us first examine ?pls and the default arguments used in the PLS function:

- ncomp = 2: The first two PLS components are calculated,
- scale = TRUE: Each data set is scaled (each variable has a variance of 1 to enable easier comparison) – data are internally centered,
- mode = regression: A PLS regression mode is performed.

As we have covered in Section 6.1, two sample plots are output: the first plot projects the X data set in the space spanned by the X−components (t_1, t_2) and the second plot projects the Y data set in the space spanned by the Y−components (u_1, u_2).

The correlation circle plot plotVar() was presented in Section 6.2. Currently, the number of X variables is too large (3116 variables) and will make the plot difficult to interpret!

10.5.2.2   sparse PLS

```
spls.result <- spls(X, Y, keepX = c(10, 20),
                    keepY = c(3, 2), ncomp = 2)    # 1 Run the method
plotIndiv(spls.result)                            # 2 Plot the samples
plotVar(spls.result)                              # 3 Plot the variables
```

In the `spls()` function, we specify the number of variables `keepX` and `keepY` to select on each component (here 2 components). The default values are the same as in PLS listed above (see `?spls`). If the argument `keepX` or `keepY` are omitted, then all variables are selected.

The correlation circle plot will appear less cluttered as it only shows the selected variables (for example, here a total of 35 variables). There are ways to tweak the aesthetic of this plot, as we will show later in this case study, along with a more in-depth analysis.

10.5.3   Example: PLS1 regression

We first start with a simple case scenario where we wish to explain one Y variable with a combination of selected X variables (transcripts). We choose the following clinical measurement which we denote as the y single response variable:

```
y <- liver.toxicity$clinic[, "ALB.g.dL."]
```

10.5.3.1   Number of dimensions using the Q^2 criterion

Defining the 'best' number of dimensions to explain the data requires we first launch a PLS1 model with a large number of components. Some of the outputs from the PLS1 object are then retrieved in the `perf()` function to calculate the Q^2 criterion using repeated 10-fold cross-validation.

```
tune.pls1.liver <- pls(X = X, Y = y, ncomp = 4, mode = 'regression')
set.seed(33)  # For reproducibility with this handbook, remove otherwise
Q2.pls1.liver <- perf(tune.pls1.liver, validation = 'Mfold',
                     folds = 10, nrepeat = 5)
plot(Q2.pls1.liver, criterion = 'Q2')
```

The plot in Figure 10.4 shows that the Q^2 value varies with the number of dimensions added to PLS1, with a decrease to negative values from two dimensions. Based on this plot we would choose only one dimension but we will still add a second dimension for the graphical outputs.

Note:

- *One dimension is not unusual given that we only include one y variable in PLS1.*

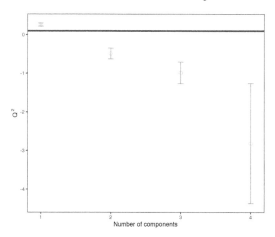

FIGURE 10.4: Q^2 **criterion to choose the number of components in PLS1.** For each dimension added to the PLS model, the Q^2 value is shown. The horizontal line of 0.0975 indicates the threshold below which adding a dimension may not be beneficial to improve accuracy in PLS.

10.5.3.2   Number of variables to select in X

We now set a grid of values – thin at the start, but also restricted to a small number of genes for a parsimonious model, which we will test for each of the two components in the `tune.spls()` function, using the MAE criterion.

```
# Set up a grid of values:
list.keepX <- c(5:10, seq(15, 50, 5))

# list.keepX  # Inspect the keepX grid
set.seed(33)  # For reproducibility with this handbook, remove otherwise
tune.spls1.MAE <- tune.spls(X, y, ncomp= 2,
                            test.keepX = list.keepX,
                            validation = 'Mfold',
                            folds = 10,
                            nrepeat = 5,
                            progressBar = FALSE,
                            measure = 'MAE')
plot(tune.spls1.MAE)
```

Figure 10.5 confirms that one dimension is sufficient to reach minimal MAE. Based on the `tune.spls()` function we extract the final parameters:

```
choice.ncomp <- tune.spls1.MAE$choice.ncomp$ncomp
# Optimal number of variables to select in X based on the MAE criterion
# We stop at choice.ncomp
choice.keepX <- tune.spls1.MAE$choice.keepX[1:choice.ncomp]

choice.ncomp
```

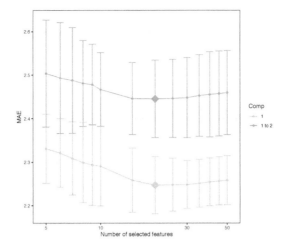

FIGURE 10.5: Mean Absolute Error criterion to choose the number of variables to select in PLS1, using repeated CV times for a grid of variables to select. The MAE increases with the addition of a second dimension (comp 1 to 2), suggesting that only one dimension is sufficient. The optimal `keepX` is indicated with a diamond.

```
## [1] 1
```

```
choice.keepX
```

```
## comp1
##    20
```

Note:

- *Other criterion could have been used and may bring different results. For example, when using measure = 'MSE', the optimal keepX was rather unstable, and is often smaller than when using the MAE criterion. As we have highlighted before, there is some back and forth in the analyses to choose the criterion and parameters that best fit our biological question and interpretation.*

10.5.3.3 Final sPLS1 model

Here is our final model with the tuned parameters:

```
spls1.liver <- spls(X, y, ncomp = choice.ncomp, keepX = choice.keepX,
                    mode = "regression")
```

The list of genes selected on component 1 can be extracted with the command line (not output here):

```
selectVar(spls1.liver, comp = 1)$X$name
```

We can compare the amount of explained variance for the X data set based on the sPLS1 (on one component) versus PLS1 (that was run on four components during the tuning step):

```
spls1.liver$prop_expl_var$X
```

```
##   comp1
## 0.08151
```

```
tune.pls1.liver$prop_expl_var$X
```

```
##   comp1   comp2   comp3   comp4
## 0.11079 0.14011 0.21715 0.06433
```

The amount of explained variance in X is lower in sPLS1 than PLS1 for the first component. However, we will see in this case study that the Mean Squared Error Prediction is also lower (better) in sPLS1 compared to PLS1.

For further graphical outputs, we need to add a second dimension in the model, which can include the same number of `keepX` variables as in the first dimension. However, the interpretation should primarily focus on the first dimension. In Figure 10.6, we colour the samples according to the time of treatment and add symbols to represent the treatment dose. Recall however that such information was not included in the sPLS1 analysis.

```
spls1.liver.c2 <- spls(X, y, ncomp = 2, keepX = c(rep(choice.keepX, 2)),
                mode = "regression")
```

```
plotIndiv(spls1.liver.c2,
          group = liver.toxicity$treatment$Time.Group,
          pch = as.factor(liver.toxicity$treatment$Dose.Group),
          legend = TRUE, legend.title = 'Time', legend.title.pch = 'Dose')
```

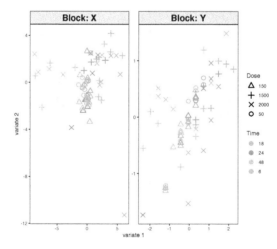

FIGURE 10.6: Sample plot from the PLS1 performed on the `liver.toxicity` data with two dimensions. Components associated to each data set (or block) are shown. Focussing only on the projection of the samples on the first component shows that the genes selected in X tend to explain the $48\,\mathrm{h}$ length of treatment vs the earlier time points. This is somewhat in agreement with the levels of the y variable. However, more insight can be obtained by plotting the first components only, as shown in Figure 10.7.

The alternative is to plot the component associated to the X data set (here corresponding to a linear combination of the selected genes) vs. the component associated to the y variable (corresponding to the scaled y variable in PLS1 with one dimension), or calculate the correlation between both components:

```
# Define factors for colours matching plotIndiv above
time.liver <- factor(liver.toxicity$treatment$Time.Group,
                levels = c('18', '24', '48', '6'))
dose.liver <- factor(liver.toxicity$treatment$Dose.Group,
                levels = c('50', '150', '1500', '2000'))
# Set up colours and symbols
col.liver <- color.mixo(time.liver)
pch.liver <- as.numeric(dose.liver)

plot(spls1.liver$variates$X, spls1.liver$variates$Y,
     xlab = 'X component', ylab = 'y component / scaled y',
     col = col.liver, pch = pch.liver)
legend('topleft', col = color.mixo(1:4), legend = levels(time.liver),
       lty = 1, title = 'Time')
legend('bottomright', legend = levels(dose.liver), pch = 1:4,
       title = 'Dose')
```

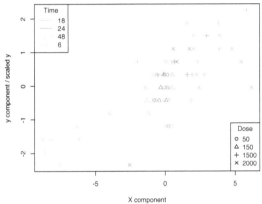

FIGURE 10.7: Sample plot from the sPLS1 performed on the `liver.toxicity` data on one dimension. A reduced representation of the 20 genes selected and combined in the X component on the x-axis with respect to the y component value (equivalent to the scaled values of y) on the y-axis. We observe a separation between the high doses 1500 and 2000 mg/kg (symbols + and ×) at 48 h and and 18 h while low and medium doses cluster in the middle of the plot. High doses for 6h and 18 h have high scores for both components.

```
cor(spls1.liver$variates$X, spls1.liver$variates$Y)
```

```
##           comp1
## comp1 0.7515
```

Figure 10.7 is a reduced representation of a multivariate regression with PLS1. It shows that

PLS1 effectively models a linear relationship between y and the combination of the 20 genes selected in X.

10.5.3.4   Performance assessment of sPLS1

The performance of the final model can be assessed with the `perf()` function, using repeated cross-validation (CV). Because a single performance value has little meaning, we propose to compare the performances of a full PLS1 model (with no variable selection) with our sPLS1 model based on the MSEP (other criteria can be used):

```
set.seed(33)   # For reproducibility with this handbook, remove otherwise

# PLS1 model and performance
pls1.liver <- pls(X, y, ncomp = choice.ncomp, mode = "regression")
perf.pls1.liver <- perf(pls1.liver, validation = "Mfold", folds =10,
                   nrepeat = 5, progressBar = FALSE)
perf.pls1.liver$measures$MSEP$summary


##    feature comp   mean      sd
## 1       Y     1 0.7282 0.04135


# To extract values across all repeats:
# perf.pls1.liver$measures$MSEP$values

# sPLS1 performance
perf.spls1.liver <- perf(spls1.liver, validation = "Mfold", folds = 10,
                   nrepeat = 5, progressBar = FALSE)
perf.spls1.liver$measures$MSEP$summary


##    feature comp   mean      sd
## 1       Y     1 0.5959 0.02698
```

The MSEP is lower with sPLS1 compared to PLS1, indicating that the X variables selected (listed above with `selectVar()`) can be considered as a good linear combination of predictors to explain y.

10.5.4   Example: PLS2 regression

PLS2 is a more complex problem than PLS1, as we are attempting to fit a linear combination of a subset of Y variables that are maximally covariant with a combination of X variables. The sparse variant allows for the selection of variables from both data sets.

As a reminder, here are the dimensions of the Y matrix that includes clinical parameters associated with liver failure.

```
dim(Y)


## [1] 64 10
```

10.5.4.1 Number of dimensions using the Q^2 criterion

Similar to PLS1, we first start by tuning the number of components to select by using the `perf()` function and the Q^2 criterion using repeated cross-validation.

```
tune.pls2.liver <- pls(X = X, Y = Y, ncomp = 5, mode = 'regression')
```

```
set.seed(33)   # For reproducibility with this handbook, remove otherwise
Q2.pls2.liver <- perf(tune.pls2.liver, validation = 'Mfold', folds = 10,
                  nrepeat = 5)
plot(Q2.pls2.liver, criterion = 'Q2.total')
```

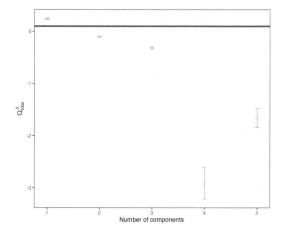

FIGURE 10.8: Q^2 criterion to choose the number of components in PLS2. For each component added to the PLS2 model, the averaged Q^2 across repeated cross-validation is shown, with the horizontal line of 0.0975 indicating the threshold below which the addition of a dimension may not be beneficial to improve accuracy.

Figure 10.8 shows that one dimension should be sufficient in PLS2. We will include a second dimension in the graphical outputs, whilst focusing our interpretation on the first dimension.

Note:

- *Here we chose repeated cross-validation, however, the conclusions were similar for* `nrepeat = 1`.

10.5.4.2 Number of variables to select in both X and Y

Using the `tune.spls()` function, we can perform repeated cross-validation to obtain some indication of the number of variables to select. We show an example of code below which may take some time to run (see `?tune.spls()` to use parallel computing). The tuning function tended to favour a very small signature, hence we decided to constrain the start of the grid to 3 for a more insightful signature. Both `measure = 'cor'` and RSS gave similar signature sizes, but this observation might differ for other case studies.

```
# This code may take several min to run, parallelisation option is possible
list.keepX <- c(seq(5, 50, 5))
list.keepY <- c(3:10)

set.seed(33)   # For reproducibility with this handbook, remove otherwise
tune.spls.liver <- tune.spls(X, Y, test.keepX = list.keepX,
                             test.keepY = list.keepY, ncomp = 2,
                             nrepeat = 1, folds = 10,
                             mode = 'regression', measure = 'cor')
```

The optimal parameters can be output, along with a plot showing the tuning results, as shown in Figure 10.9.

```
plot(tune.spls.liver)
```

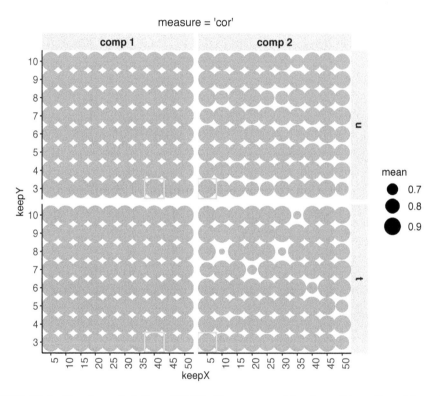

FIGURE 10.9: Tuning plot for sPLS2. For every grid value of `keepX` and `keepY`, the averaged correlation coefficients between the t and u components are shown across repeated CV, with optimal values (here corresponding to the highest mean correlation) indicated in a green square for each dimension and data set.

10.5.4.3   Final sPLS2 model

Here is our final model with the tuned parameters for our sPLS2 regression analysis. Note that if you choose to not run the tuning step, you can still decide to set the parameters of your choice here.

```
#Optimal parameters
choice.keepX <- tune.spls.liver$choice.keepX
choice.keepY <- tune.spls.liver$choice.keepY
choice.ncomp <- length(choice.keepX)

spls2.liver <- spls(X, Y, ncomp = choice.ncomp,
                    keepX = choice.keepX,
                    keepY = choice.keepY,
                    mode = "regression")
```

10.5.4.3.1 Numerical outputs

The amount of explained variance can be extracted for each dimension and each data set:

```
spls2.liver$prop_expl_var
```

```
## $X
##    comp1    comp2
## 0.19955 0.08074
##
## $Y
##   comp1   comp2
## 0.3650 0.2159
```

The selected variables can be extracted from the **selectVar()** function, for example for the X data set, with either their **$name** or the loading **$value** (not output here):

```
selectVar(spls2.liver, comp = 1)$X$value
```

The VIP measure is exported for all variables in X, here we only subset those that were selected (non null loading value) for component 1:

```
vip.spls2.liver <- vip(spls2.liver)
# just a head
head(vip.spls2.liver[selectVar(spls2.liver, comp = 1)$X$name,1])
```

```
## A_42_P620915  A_43_P14131 A_42_P578246  A_43_P11724
##        20.10        18.77        14.50        14.03
## A_42_P840776 A_42_P675890
##        13.38        12.82
```

The (full) output shows that most X variables that were selected are important for explaining Y, since their VIP is greater than 1.

We can examine how frequently each variable is selected when we subsample the data using the **perf()** function to measure how stable the signature is (Table 10.1). The same could be output for other components and the Y data set.

```
perf.spls2.liver <- perf(spls2.liver, validation = 'Mfold', folds = 10,
nrepeat = 5)
# Extract stability
stab.spls2.liver.comp1 <- perf.spls2.liver$features$stability.X$comp1
# Averaged stability of the X selected features across CV runs, as shown in
Table 10.1
stab.spls2.liver.comp1[1:choice.keepX[1]]


 # We extract the stability measures of only the variables selected in spls2.liver
extr.stab.spls2.liver.comp1 <- stab.spls2.liver.comp1[selectVar(spls2.liver,
                                                comp =1)$X$name]
```

TABLE 10.1: Stability measure (occurrence of selection) of the variables from X selected with sPLS2 across repeated 10-fold subsampling on component 1.

	X
A_42_P620915	1.00
A_43_P14131	1.00
A_42_P578246	1.00
A_43_P11724	1.00
A_42_P840776	1.00
A_42_P675890	1.00
A_42_P809565	1.00
A_43_P23376	1.00
A_43_P10606	1.00
A_43_P17415	1.00
A_42_P758454	1.00
A_42_P802628	1.00
A_43_P22616	1.00
A_42_P834104	1.00
A_42_P705413	1.00
A_42_P684538	1.00
A_43_P16842	1.00
A_43_P10003	1.00
A_42_P825290	1.00
A_42_P738559	1.00
A_43_P11570	0.92
A_42_P681650	0.98
A_42_P586270	0.92
A_43_P12400	1.00
A_42_P769476	0.94
A_42_P814010	1.00
A_42_P484423	0.96
A_42_P636498	0.90
A_43_P12806	0.88
A_43_P12832	0.86
A_42_P610788	0.72
A_42_P470649	0.86
A_43_P15425	0.72
A_42_P681533	0.86
A_42_P669630	0.66
A_43_P14864	0.60
A_42_P698740	0.56
A_42_P550264	0.52
A_43_P10006	0.42
A_42_P469551	0.34

We recommend to mainly focus on the interpretation of the most stable selected variables (with a frequency of occurrence greater than 0.8).

10.5.4.3.2   Graphical outputs

Sample plots. Using the `plotIndiv()` function, we display the sample and metadata information using the arguments `group` (colour) and `pch` (symbol) to better understand the similarities between samples modelled with sPLS2.

The plot on the left hand side corresponds to the projection of the samples from the X data set (gene expression) and the plot on the right hand side the Y data set (clinical variables).

```
plotIndiv(spls2.liver, ind.names = FALSE,
          group = liver.toxicity$treatment$Time.Group,
          pch = as.factor(liver.toxicity$treatment$Dose.Group),
          col.per.group = color.mixo(1:4),
          legend = TRUE, legend.title = 'Time',
          legend.title.pch = 'Dose')
```

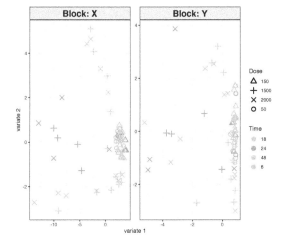

FIGURE 10.10: Sample plot for sPLS2 performed on the `liver.toxicity` data. Samples are projected into the space spanned by the components associated to each data set (or block). We observe some agreement between the data sets, and a separation of the 1500 and 2000 mg doses (+ and ×) in the 18 h, 24 h time points, and the 48 h time point.

From Figure 10.10, we observe an effect of low vs. high doses of acetaminophen (component 1) as well as time of necropsy (component 2). There is some level of agreement between the two data sets, but it is not perfect!

If you run an sPLS with three dimensions, you can consider the 3D `plotIndiv()` by specifying `style = '3d'` in the function.

The `plotArrow()` option is useful in this context to visualise the level of agreement between data sets. Recall that in this plot:

- The start of the arrow indicates the location of the sample in the X projection space,
- The end of the arrow indicates the location of the (same) sample in the Y projection space,
- Long arrows indicate a disagreement between the two projected spaces.

```
plotArrow(spls2.liver, ind.names = FALSE,
          group = liver.toxicity$treatment$Time.Group,
          col.per.group = color.mixo(1:4),
          legend.title = 'Time.Group')
```

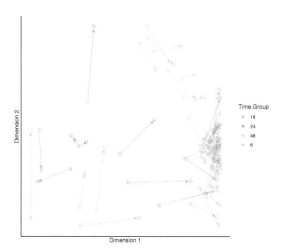

FIGURE 10.11: Arrow plot from the sPLS2 performed on the `liver.toxicity` data. The start of the arrow indicates the location of a given sample in the space spanned by the components associated to the gene data set, and the tip of the arrow the location of that same sample in the space spanned by the components associated to the clinical data set. We observe large shifts for 18 h, 24 h and 48 h samples for the high doses, however the clusters of samples remain the same, as we observed in Figure 10.10.

In Figure 10.11, we observe that specific groups of samples seem to be located far apart from one data set to the other, indicating a potential discrepancy between the information extracted. However the groups of samples according to either dose or treatment remains similar.

Variable plots. Correlation circle plots illustrate the correlation structure between the two types of variables. To display only the name of the variables from the Y data set, we use the argument `var.names = c(FALSE, TRUE)` where each Boolean indicates whether the variable names should be output for each data set. We also modify the size of the font, as shown in Figure 10.12:

```
plotVar(spls2.liver, cex = c(3,4), var.names = c(FALSE, TRUE))
```

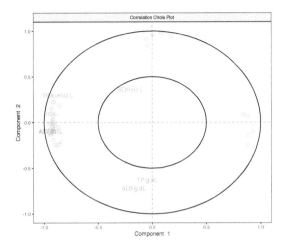

FIGURE 10.12: Correlation circle plot from the sPLS2 performed on the liver.toxicity data. The plot highlights correlations *within* selected genes (their names are not indicated here), *within* selected clinical parameters, and correlations *between* genes and clinical parameters on each dimension of sPLS2. This plot should be interpreted in relation to Figure 10.10 to better understand how the expression levels of these molecules may characterise specific sample groups.

To display variable names that are different from the original data matrix (e.g. gene ID), we set the argument `var.names` as a list for each type of label, with geneBank ID for the X data set, and `TRUE` for the Y data set (Figure 10.13):

```
plotVar(spls2.liver,
        var.names = list(X.label = liver.toxicity$gene.ID[,'geneBank'],
        Y.label = TRUE), cex = c(3,4))
```

The correlation circle plots highlight the contributing variables that, together, explain the covariance between the two data sets. In addition, specific subsets of molecules can be further investigated, and in relation to the sample group they may characterise. The latter can be examined with additional plots (for example, boxplots with respect to known sample groups and expression levels of specific variables, as we showed in the PCA case study 9.5.3). The next step would be to examine the validity of the biological relationship between the clusters of genes with some of the clinical variables that we observe in this plot.

A 3D plot is also available in `plotVar()` with the argument `style = '3d'`. It requires an sPLS2 model with at least three dimensions.

Other plots are available to complement the information from the correlation circle plots, such as Relevance networks and Clustered Image Maps (CIMs), as described in Chapter 6.

The network in sPLS2 displays only the variables selected by sPLS, with an additional `cutoff` similarity value argument (absolute value between 0 and 1) to improve interpretation. Because Rstudio sometimes struggles with the margin size of this plot, we can either launch `X11()` prior to plotting the network, or use the arguments `save` and `name.save` as shown below:

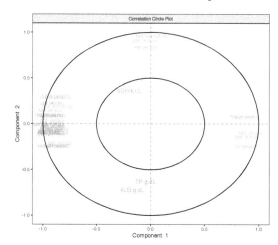

**FIGURE 10.13: Correlation circle plot from the sPLS2 performed on the
liver.toxicity data**. A variant of Figure 10.12 with gene names that are available
in $gene.ID (Note: some gene names are missing).

```
# Define red and green colours for the edges
color.edge <- color.GreenRed(50)

# X11()   # To open a new window for Rstudio
network(spls2.liver, comp = 1:2,
        cutoff = 0.7,
        shape.node = c("rectangle", "circle"),
        color.node = c("cyan", "pink"),
        color.edge = color.edge,
        # To save the plot, otherwise comment out:
        save = 'pdf', name.save = 'network_liver')
```

Figure 10.14 shows two distinct groups of variables. The first cluster groups four clinical
parameters that are mostly positively associated with selected genes. The second group
includes one clinical parameter negatively associated with other selected genes. These
observations are similar to what was observed in the correlation circle plot in Figure 10.12.

Note:

- *Whilst the edges and nodes in the network do not change, the appearance might be
  different from one run to another as the function relies on a random process to use the
  space as best as possible (using the **igraph** R package Csardi et al. (2006)).*

The Clustered Image Map also allows us to visualise correlations between variables. Here we
choose to represent the variables selected on the two dimensions and we save the plot as a
pdf figure.

```
# X11()   # To open a new window if the graphic is too large
cim(spls2.liver, comp = 1:2, xlab = "clinic", ylab = "genes",
    # To save the plot, otherwise comment out:
    save = 'pdf', name.save = 'cim_liver')
```

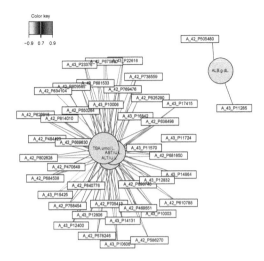

FIGURE 10.14: Network representation from the sPLS2 performed on the liver.toxicity data. The networks are bipartite, where each edge links a (rectangle) to a clinical variable (circle) node, according to a similarity matrix described in Section 6.2. Only variables selected by sPLS2 on the two dimensions are represented and are further filtered here according to a cutoff argument (optional).

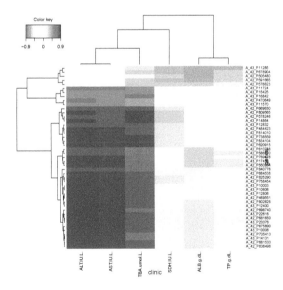

FIGURE 10.15: Clustered Image Map from the sPLS2 performed on the liver.toxicity data. The plot displays the similarity values (as described in 6.2 Section 6.2) between the X and Y variables selected across two dimensions, and clustered with a complete Euclidean distance method.

The CIM in Figure 10.15 shows that the clinical variables can be separated into three clusters, each of them either positively or negatively associated with two groups of genes. This is similar to what we have observed in Figure 10.12. We would give a similar interpretation to the relevance network, had we also used a `cutoff` threshold in `cim()`.

Note:

- *A biplot for PLS objects is also available.*

10.5.4.3.3 Performance

To finish, we can compare the performance of sPLS2 with PLS2 that includes all variables.

```
# Comparisons of final models (PLS, sPLS)

## PLS
pls.liver <- pls(X, Y, mode = 'regression', ncomp = 2)
perf.pls.liver <-  perf(pls.liver, validation = 'Mfold', folds = 10,
                   nrepeat = 5)

## Performance for the sPLS model ran earlier
perf.spls.liver <-  perf(spls2.liver, validation = 'Mfold', folds = 10,
                    nrepeat = 5)


plot(c(1,2), perf.pls.liver$measures$cor.upred$summary$mean,
     col = 'blue', pch = 16,
     ylim = c(0.6,1), xaxt = 'n',
     xlab = 'Component', ylab = 't or u Cor',
     main = 's/PLS performance based on Correlation')
axis(1, 1:2)  # X-axis label
points(perf.pls.liver$measures$cor.tpred$summary$mean, col = 'red', pch = 16)
points(perf.spls.liver$measures$cor.upred$summary$mean, col = 'blue', pch = 17)
points(perf.spls.liver$measures$cor.tpred$summary$mean, col = 'red', pch = 17)
legend('bottomleft', col = c('blue', 'red', 'blue', 'red'),
       pch = c(16, 16, 17, 17), c('u PLS', 't PLS', 'u sPLS', 't sPLS'))
```

We extract the correlation between the actual and predicted components t, u associated to each data set in Figure 10.16. The correlation remains high on the first dimension, even when variables are selected. On the second dimension the correlation coefficients are equivalent or slightly lower in sPLS compared to PLS. Overall this performance comparison indicates that the variable selection in sPLS still retains relevant information compared to a model that includes all variables.

Note:

- *Had we run a similar procedure but based on the RSS, we would have observed a lower RSS for sPLS compared to PLS.*

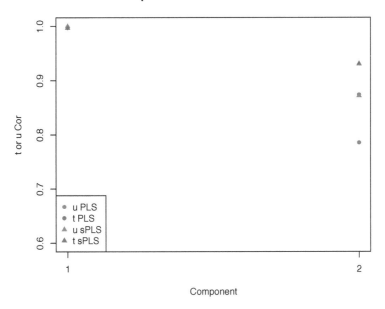

FIGURE 10.16: Comparison of the performance of PLS2 and sPLS2, based on the correlation between the actual and predicted components t, u associated to each data set for each component.

10.6 Take a detour: PLS2 regression for prediction

In this section, we describe additional analyses that can be conducted with PLS. Here we use a simple case study with the `linnerud` data from Tenenhaus (1998) to predict Y values based on X_{test} using `predict.spls()`.

```
data(linnerud)
X <- linnerud$exercise
Y <- linnerud$physiological
dim(X)
```

```
## [1] 20  3
```

```
dim(Y)
```

```
## [1] 20  3
```

As a reminder, the `linnerud` data set X contains three predictor variables; Chins, Situps, Jumps, while the Y contains three response variables; Weight, Waist, Pulse, from 20 individuals.

We will run a PLS regression model with two components (we choose `ncomp` arbitrarily here):

```
pls.linn <- pls(X, Y, ncomp = 2, mode = "regression", scale = TRUE)
```

We then artificially create two new individuals in X_{test}:

```
# New test individuals: values for Chins, Situps and Jumps
indiv1 <- c(14, 190, 35)
indiv2 <- c(2, 40, 45)

X_test <- rbind(indiv1, indiv2)
colnames(X_test) <- colnames(X)
```

Based on the PLS regression model learned on the training data, we can then predict the Y_{pred} values of these two individuals.

```
Y_pred <- predict(pls.linn, X_test)
```

The following code outputs the predicted Y_{pred} values for the first component and then for the full model that includes two components:

```
Y_pred$predict
```

```
## , , dim1
##
##        Weight Waist Pulse
## indiv1  171.5 34.20 56.94
## indiv2  196.6 38.46 53.97
##
## , , dim2
##
##        Weight Waist Pulse
## indiv1  163.2 31.96 59.08
## indiv2  201.8 39.85 52.63
```

We can represent those predicted values on the sample plot (Figure 10.17). The `predict()` function outputs the t_{pred} (the 'variates') associated to X. Thus we represent the samples in the X-space.

```
# First, plot the individuals from the learning set in the X-space
# Graphics style is needed here to overlay plots
plotIndiv(pls.linn, style = 'graphics', rep.space = 'X-variate',
          title = 'Linnerud PLS regression: sample plot')

# Then, add the test set individuals' coordinates that are predicted
points(Y_pred$variates[, 1], Y_pred$variates[, 2], pch = 17,
       col = c('red', 'blue'))
text(Y_pred$variates[, 1], Y_pred$variates[, 2], c("ind1", "ind2"),
     pos = 1, col = c('red', 'blue'))
```

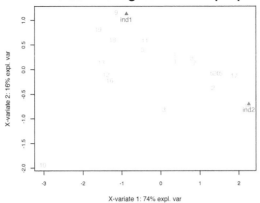

FIGURE 10.17: Predicted individual values on sample plot with PLS regression on `linnerud`. Predicted scores for individuals 1 and 2 are projected into the PLS components space of the training set.

We can look back at the training data X and investigate whether these predicted components make sense in relation to predicted individuals in X with similar measurement values. Test individual 1 is close to training individual 9. Indeed their X values are close. Similarly for test individual 2 and training individual 14. Based on these observations, we could go back to the predicted Y_{pred} values and compare with the actual values from individuals 9 and 14.

10.7 To go further

The PLS algorithm forms the basis of many methodological variants, some of which are described in this section.

10.7.1 Orthogonal projections to latent structures

O-PLS is based on a NIPALS algorithm with a PLS regression mode (Trygg and Wold, 2002). O-PLS aims to remove variation from X that is not correlated to Y (i.e. removing systematic variation in X that is orthogonal to Y). Hence, O-PLS decomposes the variance of X into correlated and orthogonal variation to the response Y, where the former can also be further analysed. The prediction of O-PLS is similar to PLS, but will yield a smaller number of dimensions. The approach is popular in the fields of chemometrics and metabolomics (Trygg and Lundstedt, 2007), and is implemented in the `ropls` package (Thévenot et al., 2015). A 'Light-sparse-OPLS' approach has recently been proposed for variable selection (Féraud et al., 2017) and is implemented in `mbxucl` R routines for metabolomics data in GitHub[2].

Note:

---

[2]https://rdrr.io/github/ManonMartin/MBXUCL/

- *Orthogonal PLS should not be confused with Orthonormalised PLS (van Gerven et al. (2012), Muñoz-Romero et al. (2015)) that is based on heteroencoders for the pattern recognition field.*

10.7.2   Redundancy analysis

Redundancy Analysis (RDA, van den Wollenberg (1977)) models directionality between data sets by fitting a linear relationship between multiple responses in the Y data set with multiple predictors from the X data set (multiple multivariable regression). RDA is solved with a two-factor PLS with a series of ordinary least squares regressions (Fornell et al., 1988), which becomes numerically difficult to solve when the number of variables is large. Sparse RDA (Csala et al., 2017) was recently proposed to select X variables to explain *all* Y variables based on elastic net penalisation (Section 3.3), and is implemented in the sRDA R package (Csala, 2017). When comparing sPLS and sRDA, we found that the variables selected in X were similar in the first dimension, but their loading weights differed. From dimension 2 the variable selection differed. The amount of explained variance in X was similar between both approaches, but sRDA maximised the variance explained in Y more efficiently (within the first component) compared to sPLS.

10.7.3   Group PLS

Taking into account the structure of groups of variables (e.g. genes that belong to the same biological pathway) can be beneficial in terms of 'augmenting' the effect of a group, but also to increase biological relevance (similar to a gene set enrichment analysis). In a context where molecules in a given data set are assigned to different pathways, group lasso enables us to select such groups (Simon and Tibshirani, 2012); (Yuan and Lin, 2006). Recent developments include sparse group lasso (Simon et al., 2013) to also select variables within each group. Both lasso variants were developed for PLS (Liquet et al., 2016) and implemented in the sgPLS R package (Liquet et al., 2017).

Note:

- *The current implementation does not allow for overlapping genes across groups, which would require additional developments.*

10.7.4   PLS path modelling

PLS-PM is a Partial Least Squares approach to Structural Equation Modeling to study complex multivariate relationships among observed and latent variables, where each data set is summarised by a latent variable (Lohmöller, 2013). PLS-PM is implemented in the plspm R package available on gitHub[3]. Whilst PLS-PM has been mostly applied in social sciences disciplines, recent areas such as engineering, environmental sciences, and medicine have used this approach to estimate complex cause-effect relationship models with latent variables. The PLS-PM framework has been extended to longitudinal models (Roemer, 2016).

Note:

---

[3]https://github.com/gastonstat/plspm

- *Other types of deflations are available in PLS and in our package. The `invariant` mode performs a redundancy analysis, where the Y matrix is not deflated. The `classic` mode is similar to a regression mode: it gives identical results for the variates and loadings associated to the X data set, but different loading vectors associated to the Y data set (since different normalisations are used). Classic mode is the PLS2 model as defined by Tenenhaus (1998). We have not focused our implementation and development efforts on those modes.*

10.7.5 Other sPLS variants

Several other sparse PLS methods are closely related, and were proposed by Waaijenborg et al. (2008), Witten et al. (2009), Parkhomenko et al. (2009), and Chun and Keleş (2010). Whilst some of these approaches may refer to sparse Canonical Correlation Analysis (covered in Chapter 11), they are, in fact, based on the PLS algorithm with a canonical mode. Waaijenborg et al. (2008) uses an elastic net penalty, whereas Parkhomenko et al. (2009) and Witten et al. (2009) use a lasso penalty. Our sPLS approach is similar to Parkhomenko et al. (2009) in terms of formulation (with a Lagrange form of the constraints), whereas Witten et al. (2009) proposes a bound form.

The tuning criteria also differ, based on permutations (Witten et al., 2009), MSEP (Chun and Keleş, 2010), maximisation of the correlation between components on the test set using cross-validation (Parkhomenko et al., 2009), and finding the minimum mean absolute difference between the correlation of the components from the training set, and the correlation of the components from the test set (Waaijenborg et al., 2008). According to Wilms and Croux (2015), these tuning approaches tend to result in very small variable selections. Two R packages are available, spls (Chung et al., 2020) and PMA (Witten and Tibshirani, 2020), but none offer graphical visualisations.

10.8 FAQ

Can PLS handle missing values?

Yes, but for prediction/cross-validation with `perf()` or `tune()`, we can only allow for missing values in the X data set in PLS regression mode.

Can I integrate more than two data sets with PLS?

sPLS can only manage two data sets, but see Chapter 13 for multi-block analyses.

What are the differences between sPLS canonical mode, and Canonical Correlation Analysis (CCA, see ?rcca in mixOmics)?

CCA maximises the correlation between components, as we further detail in Chapter 11; PLS maximises the covariance. Both methods give similar results if the components are scaled, but the underlying algorithms are different:

- CCA calculates all components at once, there is no deflation step involved,
- PLS has different deflation modes,
- sparse PLS selects variables, CCA cannot perform variable selection.

Can I perform PLS with more variables than observations?

Yes, and sparse PLS is particularly useful for identifying sets of variables that play a role in explaining the relationship between two data sets.

Can I perform PLS with two data sets that are highly unbalanced (thousands of variables in one data set and less than 10 in the other)?

Yes! Even if you performed sPLS to select variables in one data set (or both), you can still control the number of variables selected with `keepX`.

What about repeated measurements? The multilevel data decomposition described in Appendix 4.A can be applied here externally, using the `withinVariation()` function to extract first the within variance data that is then input into PLS.

What about compositional data? Data can be transformed with Centered Log Ratio using the `logratio.transfo()` function first (you may need to specify an `offset` value if your data contain zeros), as described in Section 4.1. The CLR data can then be input into PLS.

What about longitudinal experiments? Analysing longitudinal experiments include asking ourselves many different types of questions, including:

- Can we observe global time differences between sample groups in each omics?

- Can we observe coordinated changes in expression levels across time and omics, within and between sample groups?

Some of those questions can be answered using the `timeOmics` sister package (Bodein et al., 2020) based on smoothing splines that are included in sPLS (Straube et al., 2015); (Bodein et al., 2019). However, these methods are still in development.

10.9   Summary

PLS enables the integration of two data matrices measured on the same samples. Both supervised (regression) and unsupervised approaches are available. PLS reduces the dimensions of both data sets while performing alternate multivariate regressions. In particular, it aims to incorporate information from *both* data sets via components and loading vectors to maximise the covariance between the data sets. PLS is able to manage many noisy, collinear (correlated) and missing variables while also simultaneously modelling several response variables. The PLS algorithm is flexible and can therefore be used to answer a variety of biological questions, such as predicting the abundance of a protein given the expression of a small subset of genes, or selecting a subset of correlated biological molecules (e.g. proteins and transcripts) from matching data sets. sparse PLS further improves biological interpretability by including penalisations in the PLS method on each pair of loading vectors to select a subset of co-variant variables within the two data sets. The optimal choice of parameters in both PLS2 and sPLS2 (multivariate regression) can be difficult and subject to interpretation.

10.A Appendix: PLS algorithm

10.A.1 PLS Pseudo algorithm

We describe here the pseudo algorithm summarised in Section 10.2.

PLS2 algorithm

INITIALISE the first set of components (t, u) as the first column of X and Y, respectively[4]
  FOR EACH dimension h
   UNTIL CONVERGENCE of a
    Calculate the vectors associated to X:
     (a) Loading vector $a = X^T u / u' u$ and norm a to 1
     (b) Component $t = X a / a' a$
    Calculate the vectors associated to Y:
     (c) Loading vector $b = Y^T t / t' t$ and norm b to 1
     (d) Component $u = Y b / b' b$
   END CONVERGENCE
  Calculate the regression coefficients:
   $c = X^T t / t' t$ related to the information from X
   $d = Y^T t / t' t \propto b$
   $e = Y^T u / u' u$ related to the information from Y
  DEFLATION:
   $-X = X - tc'$
   – Regression mode: $Y = Y - td'$
   – Canonical mode: $Y = Y - ue'$
   – Increment to dimension $h + 1$

The steps (a) and (c) are the local regressions of the matrices X and Y on u and t, respectively. The components are calculated in steps (b) and (d) and can be seen as regressing the matrices on their respective loading vectors, as we represent in Figure 10.18.

Note:

- *Even if a and b are normed, the division by $a' a$ and $b' b$ in steps (b) and (d) enable us to manage missing values.*

The calculation of the regression coefficient d looks very similar to the calculation of the loading vector b except that in step (c), b has a ℓ_2 norm of 1 (effectively calculated as $b_{\text{normed}} = \frac{b}{||b||_2}$, so that $||b_{\text{normed}}||_2 = 1$). Thus d is proportional to b up to a constant.

For the deflation, we calculate the regression coefficients d, c and e. The latter two are obtained by regressing the current matrices X and Y on the component u.

---

[4]A faster alternative is to obtain the loading vectors (a, b) from $\text{SVD}(X^T Y)$, then update (t, u) in steps (b) and (d).

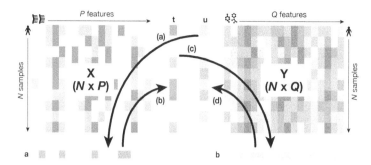

FIGURE 10.18: Schematic view of the alternate local regressions in the PLS algorithm. The end of the arrows indicate the resulting vector (either latent variable or loading vector) and the letters indicate the steps in the PLS algorithm for the first dimension.

Notes:

- *The PLS variant denoted* `mode` *=* `classic` *deflates* $\boldsymbol{Y}$ *with respect to* $\boldsymbol{b}$ *rather than* $\boldsymbol{d}$ *(see Tenenhaus (1998)), and is also implemented in* `mixOmics` *but we do not use this type of deflation in practice).*

- *If* $\boldsymbol{Y}$ *includes only one variable, then we obtain a PLS1.*

10.A.2   Convergence of the PLS iterative algorithm

For a given dimension h, after the convergence step, the vectors $\boldsymbol{a}^h$, $\boldsymbol{b}^h$, $\boldsymbol{t}^h$, and $\boldsymbol{u}^h$ verify the equations:

$$
\begin{aligned}
\boldsymbol{X}^{(h-1)T}\boldsymbol{Y}^{(h-1)}\boldsymbol{Y}^{(h-1)T}\boldsymbol{X}^{(h-1)}\boldsymbol{a}^h &= \delta_h \boldsymbol{a}^h \\
\boldsymbol{X}^{(h-1)}\boldsymbol{X}^{(h-1)T}\boldsymbol{Y}^{(h-1)}\boldsymbol{Y}^{(h-1)T}\boldsymbol{t}^h &= \delta_h \boldsymbol{t}^h \\
\boldsymbol{Y}^{(h-1)T}\boldsymbol{X}^{(h-1)}\boldsymbol{X}^{(h-1)T}\boldsymbol{Y}^{(h-1)}\boldsymbol{b}^h &= \delta_h \boldsymbol{b}^h \\
\boldsymbol{Y}^{(h-1)}\boldsymbol{Y}^{(h-1)T}\boldsymbol{X}^{(h-1)}\boldsymbol{X}^{(h-1)T}\boldsymbol{u}^h &= \delta_h \boldsymbol{u}^h
\end{aligned}
\tag{10.3}
$$

The PLS Pseudo algorithm above solves an eigenvalue problem with the iterative power method. The resolution of one of these equations is sufficient to solve the whole equation system by computing the other eigenvectors. According to Hoskuldsson (1988), the PLS2 algorithm should converge in less than 10 iterations.

10.A.3   PLS-SVD method

The PLS-SVD variant solves the PLS objective function in a more efficient manner by decomposing the $\boldsymbol{X}^T\boldsymbol{Y}$ matrix into singular values and vectors (Lorber et al., 1987). The SVD decomposition is a generalisation of the eigen decomposition problem (Golub and Van Loan, 1996). We have seen that using the SVD provides the solution to PCA, but also to other multivariate analyses, such as CCA and Linear Discriminant Analysis that use generalised SVD (Jolliffe, 2005).

For PLS, we use a classical SVD decomposition of $X^T Y$, which can be written as:

$$X^T Y = A \Delta B^T \tag{10.4}$$

where A ($P \times r$) and B ($Q \times r$) are the matrices containing left and right singular vectors (a^h and b^h) and correspond to the PLS loading vectors, $h = 1, \ldots, H$, $H \leq r$, where r is the rank of the matrix $X^T Y$. The singular values in Δ are the eigenvalues δ_h in the cyclic equations (10.3).

The deflation step uses the rank-1 approximation of the $X^T Y$ matrix up to dimension h as:

$$(X^T Y)^{(h)} = (X^T Y)^{(h-1)} - \sum_{k=1}^{h-1} \delta_k a^k b^{k'}. \tag{10.5}$$

This algorithm avoids computing the iterative power method and is computationally efficient, as the SVD decomposition is computed only once to obtain all the loading component vectors. There might be a risk of possible loss of orthogonality between the latent variables t^h and u^h, although we have not observed this phenomenon in practice.

10.B Appendix: sparse PLS

10.B.1 sparse PLS-SVD

Using the framework of Shen and Huang (2008) who proposed a sparse PCA with lasso penalty, we implemented a sparse PLS by penalising simultaneously the loading vectors from X and Y (Lê Cao et al., 2008). Denoting $M = X^T Y$, the optimisation problem becomes:

$$\min_{a,b} ||M - ab'||_F^2 + P_{\lambda_1}(a) + P_{\lambda_2}(b), \tag{10.6}$$

where $P_\lambda(.)$ is the soft thresholding function with the penalty λ that is applied on each element of the vectors a and b (as defined in Equation (9.6)). The lasso penalties λ_1 and λ_2 do not need to be equal.

For the current M matrix, we penalise the loading vectors in the iterative PLS algorithm as:

$$b_{new} = P_{\lambda_1}(M a_{old})$$
$$a_{new} = P_{\lambda_2}(M^T b_{old}).$$

During the deflation step of the PLS-SVD, since $M^{(h)} \neq X^{(h)T} Y^{(h)}$ (see Equation (10.5)), we separately compute $X^{(h)}$ and $Y^{(h)}$, and define $\tilde{M}^{(h)} = X^{(h)T} Y^{(h)}$ for each step to extract the first pair of singular (loadings) vectors a and b that are then input in the PLS algorithm. Considering the first pair of loading vectors at a time will lead to the biggest reduction of the total variation in the X- and Y-spaces (Hoskuldsson, 1988). In our approach, the SVD decomposition framework provides a useful tool for selecting variables from each data set.

10.B.2 sparse PLS pseudo algorithm

We detail the sparse PLS algorithm (sPLS) based on the iterative PLS algorithm (Tenenhaus, 1998) and SVD computation of $\tilde{M}^{(h)}$ for each dimension.

sPLS algorithm

INITIALISE for each dimension h
   – $\tilde{M} = X^T Y$
   – SVD($\tilde{M}$) to extract the first pair of singular vectors a and b
   UNTIL CONVERGENCE of a:
        (a) $a = P_{\lambda_2}(\tilde{M}b)$, norm a to 1
        (b) $b = P_{\lambda_1}(\tilde{M}^T a)$, norm b to 1
   – Calculate the components $t = Xa$ and $u = Yb$
   END CONVERGENCE
   Calculate the regression coefficients:
   – $c = X^T t / t't$
   – $d = Y^T t / t't$
   – $e = Y^T u / u'u$
   DEFLATION
 – $X = X - tc'$
 – Regression mode: $Y = Y - td'$
 – Canonical mode: $Y = Y - te'$
 – Increment to next dimension $h + 1$
END DIMENSION

In the case where there is no sparsity constraint ($\lambda_1 = \lambda_2 = 0$), this approach is equivalent to a PLS with no variable selection.

10.C   Appendix: Tuning the number of components

We detail how to calculate the Q^2 to choose the number of components in PLS1 and PLS2 from Section 10.3.

10.C.1   In PLS1

10.C.1.1   Fitted values for PLS regression mode

We actually have used the fitted $\hat{Y}$ in the deflation step in the PLS pseudo algorithm above to calculate the residual matrix. For a regression mode, we have:

$$Y^{(h+1)} = Y^{(h)} - \hat{Y} = Y^{(h)} - td'$$

The fitted values $\hat{Y} = td'$ are calculated based on the reconstruction of Y using the component t and the regression coefficient vector d for a given dimension h[5].

Here we consider the fitted $\hat{Y}_{ij}$ on all samples i, and variables j. We define the Residual Sum of Squares RSS_{hj} for a given dimension h and for all j variables, $j = 1 \ldots, Q$ as:

$$\text{RSS}_{hj} = \sum_{i=1}^{N} (Y_{ij}^{(h)} - \hat{Y}_{ij})^2 = \sum_{i=1}^{N} (Y_{ij}^{(h+1)})^2$$

---

[5] For the sake of simplicity here, we are not using the subscripts ^{h} that would reflect that the vectors are associated to dimension h.

For PLS1, we denote $\boldsymbol{Y}_i = \boldsymbol{y}_i$, since it is a single variable, and

$$\text{RSS}_h = \sum_{i=1}^{N} (\boldsymbol{y}_i^{(h)} - \hat{\boldsymbol{y}}_i)^2 \tag{10.7}$$

For PLS2, we sum RSS_{hj} across all Q variables:

$$\text{RSS}_h = \sum_{j=1}^{Q} \text{RSS}_{hj} \tag{10.8}$$

10.C.1.2 Predicted values for PLS regression mode

Prediction implies we have test samples defined, for example, by using cross-validation. To predict $\hat{\boldsymbol{Y}}_{\text{test}}$, we use the similar reconstruction as above, but calculated now on the test set from $\boldsymbol{X}_{\text{test}}$:

$$\hat{\boldsymbol{Y}}_{\text{test}} = \boldsymbol{t}_{\text{test}}\boldsymbol{d}' = \boldsymbol{X}_{\text{test}}\boldsymbol{a}\boldsymbol{d}'$$

where $\boldsymbol{a}$ and $\boldsymbol{d}$ are calculated on the training set. $\boldsymbol{X}_{\text{test}}\boldsymbol{a}$ is the predicted component $\boldsymbol{t}_{\text{test}}$, as we have seen already when tuning PCA in Section 9.B.

We define the PREdicted Sum of Squares PRESS_{hj} for a given dimension h and for all j variables, $j = 1 \ldots, Q$ as:

$$\text{PRESS}_{hj} = \sum_{i \in \text{test}} (\boldsymbol{Y}_{ij}^{(h)} - \hat{\boldsymbol{Y}}_{ij})^2$$

For PLS1, using the notations as above,

$$\text{PRESS}_h = \sum_{i \in \text{test}} (\boldsymbol{y}_i^{(h)} - \hat{\boldsymbol{y}}_i)^2. \tag{10.9}$$

For PLS2, we then sum PRESS_{hj} across all Q variables:

$$\text{PRESS}_h = \sum_{j=1}^{Q} \text{PRESS}_{hj} \tag{10.10}$$

10.C.1.3 Fitted and predicted values for PLS canonical mode

For this mode, we focus instead on the fitted and predictive values $\hat{\boldsymbol{X}}$ and $\hat{\boldsymbol{X}}_{\text{test}}$.

The fitted values are defined as:

$$\hat{\boldsymbol{X}} = \boldsymbol{t}\boldsymbol{c}'$$

calculated based on the reconstruction of $\boldsymbol{X}$ using the component $\boldsymbol{t}$ and the regression coefficient vector $\boldsymbol{c}$ for a given dimension h.

The Residual Sum of Squares RSS_h for a given dimension h and across all variables from $\boldsymbol{X}$, $j = 1 \ldots, P$ is:

$$\text{RSS}_h = \sum_{j=1}^{P} \sum_{i=1}^{N} (\boldsymbol{X}_{ij}^{(h)} - \hat{\boldsymbol{X}}_{ij})^2.$$

For the predicted values $\hat{\boldsymbol{X}}_{\text{test}}$ is defined as:

$$\hat{\boldsymbol{X}}_{\text{test}} = \boldsymbol{t}_{\text{test}} c' = \boldsymbol{X}_{\text{test}} \boldsymbol{a} c',$$

where $\boldsymbol{a}$ and $\boldsymbol{c}$ are calculated on the training set.

The PREdicted Sum of Squares PRESS_h for a given dimension h across all variables in $\boldsymbol{X}$ is defined as:

$$\text{PRESS}_h = \sum_{j=1}^{P} \sum_{i \in \text{test}} (\boldsymbol{X}_{ij}^{(h)} - \hat{\boldsymbol{X}}_{ij})^2$$

10.C.1.4 Q^2 criterion

As mentioned in Section 10.3, the Q^2 assesses the gain in prediction accuracy when a dimension is added to the PLS model, by comparing the predicted values for dimension h, PRESS_h to the fitted values from the previous dimension h, RSS_{h-1}. Thus, we can decide to only include a new dimension h if:

$$\sqrt{\text{PRESS}_h} \leq 0.95 \sqrt{\text{RSS}_{h-1}}, \qquad (10.11)$$

the value of 0.95 being chosen arbitrarily here (as in the SIMCA-P software (Umetri, 1996)).

The Q^2 criterion is defined for each dimension h as:

$$Q_h^2 = 1 - \frac{\text{PRESS}_h}{\text{RSS}_{h-1}} \qquad (10.12)$$

using the equations (10.7) and (10.9) for PLS1 and (10.8) and (10.10) for PLS2. By including (10.12) and rearranging the terms in (10.11), the rule to decide whether to add dimension h in a PLS model is if:

$$Q_h^2 \geq (1 - 0.95^2) = 0.0975$$

meaning that we should stop adding new dimensions to the model when $Q_h^2 < 0.0975$.

Notes:

- *For $h = 1$, denote $\bar{\boldsymbol{Y}}_{ij}$ the sample mean of $\boldsymbol{Y}_{ij}$, we set $RSS_0 = \sum_{i=1}^{N} (\boldsymbol{Y}_{ij}^{(h)} - \bar{\boldsymbol{Y}}_{ij})^2$ in Equation (10.12). This corresponds to a naïve way of fitting the values by replacing them by their sample means at the start of the procedure.*

- $Q_h^2 \leq 0$ *indicates a poor fit.*

10.C.2   In PLS2

The criterion we use extends what was presented for sPCA in Section 9.B but for the PLS2 algorithm presented earlier.

The predicted components are defined as:

$$t_{\text{test}} = X_{\text{test}} a$$

and

$$u_{\text{test}} = Y_{\text{test}} b,$$

where a and b are obtained from the training set.

We can compare the predicted and actual components (t_{test}, t) and (u_{test}, u) based on either their correlation (in absolute value) or their sum of squared difference, which we define as

$$\text{RSS}_t = \frac{1}{N-1} \sum_{i=1}^{N} (t_{\text{test}i} - t_i)^2$$

(and similarly for RSS_u based on u).

We choose the (`keepX`, `keepY`) combination that either maximises the correlation, or minimises the RSS.

For a regression mode, both (`keepX`, `keepY`) values are determined based on (u_{test}, u) since we are interested in explaining as best as we can the Y matrix. For a canonical mode where the focus is also to explain the X matrix, we choose a `keepX` that is based on (t_{test}, t) and a `keepY` that is based on (u_{test}, u).

11

Canonical Correlation Analysis (CCA)

11.1 Why use CCA?

Canonical Correlation Analysis (CCA) (Hotelling, 1936) assesses whether two omics data sets measured on the same samples contain similar information by maximising their correlation via canonical variates. In this approach, both data sets play a *symmetric* role. This Chapter presents two variants of CCA for small and larger data sets. CCA is an exploratory approach, and we illustrate the main graphical outputs on the `nutrimouse` study.

11.1.1 Biological question

I have two omics data sets measured on the same samples. Does the information from both data sets agree and reflect any biological condition of interest? What is the overall correlation between the two data sets?

11.1.2 Statistical point of view

The biological question suggests an integration framework based on correlation, set in an unsupervised manner. In the same vein as PLS, CCA seeks for linear combinations of the variables (called 'canonical variates') to reduce the dimensions of the data sets. However, instead of simultaneously maximising the covariance between data sets, as in PLS, CCA seeks to maximise the *correlation* (called 'canonical correlation') between the pair of canonical variates. We consider two types of approaches:

- Classical Canonical Correlation Analysis (CCA) when the total number of variables $P + Q$ is less than the number of samples N and when the variables in each data set are not highly correlated (non collinear).

- regularised Canonical Correlation Analysis (rCCA) (Leurgans et al., 1993, González et al. (2008)), a variant of CCA when $N \ll P + Q$ or when the variables are highly collinear.

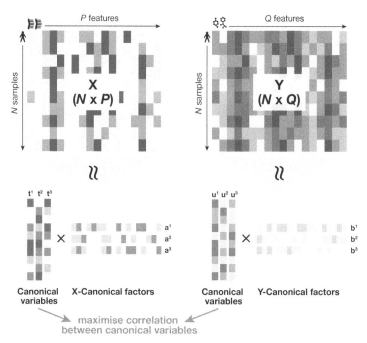

FIGURE 11.1: Schematic view of the CCA matrix decomposition of X and Y into sets of canonical variates and loading vectors. The canonical factors are defined so that the correlation between canonical variates is maximised. The canonical factors are not directly interpretable as in PLS, but associations can be interpreted via graphical outputs, as illustrated in this chapter.

11.2   Principle

We use the following notations: X is a $(N \times P)$ data matrix, and Y is an $(N \times Q)$ data matrix, with N denoting the number of samples (individuals), and P and Q the number of variables in X and Y respectively. Samples match between X and Y.

11.2.1   CCA

CCA maximises the correlation between pairs of canonical variates (t, u) which are linear combinations of variables from X and Y respectively (see details in Appendix 11.A and Figure 11.1).

For the first dimension, CCA maximises the first canonical correlation $\rho_1 = \text{cor}(t^1, u^1)$. The objective function is:

$$\underset{var(t^1)=var(u^1)=1}{\text{argmax}} \quad \text{cor}(t^1, u^1) \tag{11.1}$$

For the second dimension, CCA maximises the second canonical correlation: $\rho_2 = \text{cor}(t^2, u^2)$ with an additional constraint:

$$\underset{var(\boldsymbol{t}^2)=var(\boldsymbol{u}^2)=1}{\text{argmax}} \quad cor(\boldsymbol{t}^2, \boldsymbol{u}^2) \quad \text{such that} \quad cor(\boldsymbol{t}^1, \boldsymbol{t}^2) = cor(\boldsymbol{u}^1, \boldsymbol{u}^2) = 0 \qquad (11.2)$$

When we consider several dimensions, the size of the canonical correlations should decrease: $\rho_1 \geq \rho_2 \geq ... \geq \rho_H$.

The main limitation of CCA occurs when the number of variables is large. To solve CCA, we need to calculate the inverse of the variance-covariance matrix $\boldsymbol{X}^T \boldsymbol{Y}$ (introduced in 3.1) that is ill-conditioned as many variables are correlated within each data set. This leads to unreliable matrix inverse estimates and results in several canonical correlations that are close to 1. In that case, the canonical subspace cannot uncover any meaningful observations. One way to handle the high dimensionality is to include a regularisation step in the CCA calculation as described below with regularised CCA.

11.2.2 rCCA

The ill-posed matrix problem in the context of high dimensional data can be by-passed by introducing constants, also called ridge penalties (see Section 3.3), on the diagonal of the variance-covariance matrices of $\boldsymbol{X}$ and $\boldsymbol{Y}$ to make them invertible. It is then possible to perform classical CCA, substituting with these regularised covariance matrices.

11.3 Input arguments and tuning

In classic CCA, the data set with the smaller number of variables is set as $\boldsymbol{X}$. In rCCA, the input order of the data matrices does not matter.

Missing values are handled internally in the function `rcc()` by setting the NA values to zero when calculating the canonical variates. However, when using the shrinkage approach described below, missing values need to be imputed beforehand to estimate the regularisation parameters (for example with NIPALS, see Section 9.5).

Since CCA and rCCA focus on maximising data set correlation (defined in Section 3.3), it does not matter whether the data are centered and scaled and hence, the scaling argument does not appear in the function.

11.3.1 CCA

The parameter to tune in CCA is the number of dimensions H to highlight relevant associations between the data sets. Similarly to PCA, there is no straightforward criterion, but a small H is preferred for interpretation. We expect that a small number of dimensions should be able to highlight the correlation structure between the two data sets in this exploratory approach.

The canonical correlations decrease in size as H increases: after a while, no relevant linear relationships are extracted between the $\boldsymbol{X}$ and the $\boldsymbol{Y}$ variables.

To choose H, we rely on graphical representations such as the screeplot (barplot of the canonical correlations vs. the dimensions) and correlation circle plots. For the latter, if most of the variables appear within the small circle of radius 0.5, it indicates that these dimensions are not relevant to extract common information between data sets.

CCA assumes variables within data sets are not correlated. This is unlikely to be the case, and especially when $P + Q >> N$. In that case rCCA is required.

11.3.2 rCCA

In rCCA, the parameters (λ_1, λ_2) are used to regularise the correlation matrices to solve CCA. They can be estimated using two different types of methods:

1. **Cross-validation (CV) approach**: Choosing regularisation parameters (λ_1, λ_2) that maximise the canonical correlation using cross-validation has been proposed (González et al., 2008). In the `tune.rcc()` function, a coarse grid of possible values for λ_1 and λ_2 is input to assess every possible pair of parameters. The default grid is defined with five equally-spaced discretisation points of the interval [0.001, 1] for each dimension, but this can be changed. A thinner grid within the coarse grid can then be refined as a second tuning step. As this process is computationally intensive, it may not run for very large data sets (P or Q > 5,000). This approach has been shown to give biologically relevant results (Combes et al., 2008). The tuning function outputs the optimal regularisation parameters, which are then input into the `rcc()` function with the argument `method = 'ridge'`, see further details in Appendix 11.A.

2. **Shrinkage approach**: This effective approach proposes an analytical calculation of the regularisation parameters for large-scale correlation matrices (Schäfer and Strimmer, 2005), and is implemented directly in the `rcc()` function using the argument `method = 'shrinkage'`. The downside of this approach is that the (λ_1, λ_2) values are calculated independently, regardless of the cross-correlation between X and Y, and thus may not be successful in optimising the correlation between the data sets. See further details in Appendix 11.A.

Both approaches have advantages and disadvantages. In the case study in Section 11.5 we apply and compare both tuning strategies in detail.

11.4 Key outputs

11.4.1 Graphical outputs

11.4.1.1 Correlation circle plots

In contrast to PCA and PLS, the canonical factors cannot be directly interpreted to identify the importance of the variables when relating X and Y (Gittins, 1985; Tenenhaus, 1998). However, a graphical representation of the variables on a correlation circle plot is feasible by projecting X and Y onto the spaces spanned by the canonical variates t and u for each dimension (see Section 6.2). The variable coordinates are the correlation coefficients between the initial variables and the canonical variates. Such representation, as we have covered previously, enables the identification of groups of variables that are correlated between the two data sets (González et al., 2012). Since CCA and rCCA do not perform variable selection, the number of variables ($= P + Q$) might be too large for interpretation. We recommend using the argument `cutoff` to hide variables that are close to the origin of the plot.

11.4.1.2 Sample plots

As we have mentioned in Section 6.1, two sample plots are output: the first plot projects the X data set in the space spanned by the X-variates (t_1, t_2) and the second plot projects the Y data set in the space spanned by the Y-variates (u_1, u_2). When using the argument `rep.space = "XY-variate"` the variates are averaged and only one plot is shown. This makes sense in the CCA context, since the correlation between variates should be maximised (and hence very similar). One way to inspect this is to use the `plotArrow()` plot first, as we illustrate in the case study.

11.4.1.3 Relevance networks and Clustered Image Maps

The same similarity matrix is input into `network()` and `cim()`, as we detailed in Section 6.2 and Appendix 6.A. You may wish to specify the dimensions with the argument `comp`, that is set by default to `1:H`, where H is the chosen dimension of the (r)CCA model.

11.4.2 Numerical outputs

The numerical outputs of interest in rCCA include:

- The H canonical correlations (per dimension).
- The regularisation parameters (λ_1, λ_2) – this is especially useful with the shrinkage method to clarify the estimation of those parameters.

11.5 Case study: Nutrimouse

The `nutrimouse` data set contains the expression levels of genes potentially involved in nutritional problems and the concentrations of hepatic fatty acids for forty mice. The data sets come from a nutrigenomic study from Martin et al. (2007), which considered the effects of five dietary regimens with contrasted fatty acid compositions on liver lipids and hepatic gene expression in mice. Two sets of variables were acquired:

- `$gene`: Expression levels of 120 genes from liver cells, selected from approximately 30,000 genes, as potentially relevant in the context of the nutrition study. These expressions come from a nylon microarray with radioactive labelling.

- `$lipid`: Concentrations of 21 hepatic fatty acids measured by gas chromatography.

Biological units (mice) were cross-classified with a two-factor experimental design (four replicates):

- `$genotype`: two-levels factor: wild-type (WT) and PPAR $-/-$ (PPAR).

- $diet: five-levels factor: Mice were fed either corn and colza oils (50/50) for a reference diet (REF), hydrogenated coconut oil for a saturated fatty acid diet (COC), sunflower oil for an Omega 6 fatty acid-rich diet (SUN), linseed oil for an Omega 3-rich diet (LIN) and corn/colza/enriched fish oils for the FISH diet (43/43/14).

11.5.1   Load the data

```
library(mixOmics)
data(nutrimouse)
X <- nutrimouse$gene
Y <- nutrimouse$lipid
```

11.5.2   Quick start

Here are the key functions for a quick start, using the functions cca(), rcca() and the graphic visualisations plotIndiv() and plotVar().

11.5.2.1   CCA

We provide below the quick start code for CCA as an example. However, this code will not run on the **nutrimouse** study as the data sets include a too large number of variables. We will use rCCA later in this case study to analyse these data.

```
result.cca.nutrimouse <- rcc(Y, X)      # 1 Run the method
plotIndiv(result.cca.nutrimouse)        # 2 Plot the samples
plotVar(result.cca.nutrimouse)          # 3 Plot the variables
```

Let us first examine ?rcc and the default arguments used in this function:

- ncomp = 2: The first two CCA dimensions are calculated,
- method = c("ridge", "shrinkage"): The method for the regularisation parameters (not needed for classical CCA),
- lambda1 = 0, lambda2 = 0: The regularisation parameters (not needed for classical CCA).

Note that there are no center nor scale options as we discussed previously. We have swapped the X and Y data sets in the call of this function to fit a classical CCA framework (where the first data set should be the one with the smaller number of variables).

By default the regularisation parameters are set to 0, meaning that a classic CCA will be performed. In the case where $P + Q > N$ you should observe in the resulting object that the canonical correlations are very close to 1 ($cor output). This is a strong indication that regularised CCA is needed (see next).

11.5.2.2 rCCA

11.5.2.2.1 With the ridge regression method

Here we will need to input the regularisation parameters (that will be tuned with `tune.rcc()` using cross-validation, as we describe later), for example:

```
result.rcca.nutrimouse <- rcc(X, Y, method = 'ridge', lambda1 = 0.5,
                              lambda2 = 0.05)      # 1 Run the method
```

11.5.2.2.2 With the shrinkage estimation method

Here there is no need to input the regularisation parameters as they are directly estimated in the function:

```
result.rcca.nutrimouse <- rcc(X, Y, method = 'shrinkage') # 1 Run the method
```

The remainder of the key graphical outputs are as usual:

```
plotIndiv(result.rcca.nutrimouse)          # 2 Plot the samples
plotVar(result.rcca.nutrimouse)            # 3 Plot the variables
```

The correlation circle plot will still appear cluttered, as rCCA does not perform variable selection. A `cutoff` argument in `plotVar()` is recommended for easier interpretation.

11.5.3 Example: CCA

We first perform a classical CCA on a smaller subset of data that include 10 genes and 10 lipids from **nutrimouse**. In a second analysis, we then consider a slightly larger number of variables (21 lipids and 25 genes) to compare the canonical correlations for both case scenarios.

```
data(nutrimouse)
X.subset1 <- nutrimouse$lipid[, 1:10]
Y.subset1 <- nutrimouse$gene[, 1:10]
subset1.rcc.nutrimouse <- rcc(X.subset1, Y.subset1)


X.subset2 <- nutrimouse$lipid    # All 21 lipids
Y.subset2 <- nutrimouse$gene[, 1:25]
subset2.rcc.nutrimouse <- rcc(X.subset2, Y.subset2)
```

By default CCA will output the canonical correlations for all possible dimensions (equal to the lowest rank of the smaller data matrix), as shown in Figure 11.2):

```
plot(subset1.rcc.nutrimouse, type = "barplot")
```

```
plot(subset2.rcc.nutrimouse, type = "barplot")
```

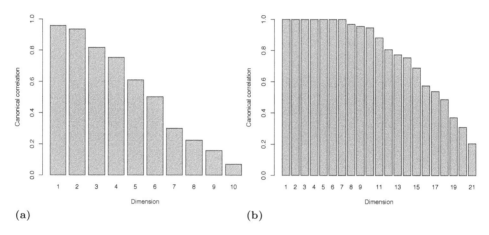

(a)   (b)

FIGURE 11.2: Classical CCA for a subset of the nutrimouse study: canonical correlations screeplot. The plot shows the canonical correlations on the y-axis with respect to the CCA dimension on the x-axis, for all possible CCA dimensions. **(a)** When we consider $P = 10$ and $Q = 10$ (i.e. $N > P + Q$); **(b)** When we consider $P = 21$ and $Q = 25$ (i.e. $N << P + Q$). Canonical correlations tend to be equal or very close to 1 when the number of variables becomes large.

The screeplots clearly show the limitation of CCA when the total number of variables is larger than the number of samples: the canonical correlations do not decrease sharply with the dimensions, and thus dimension reduction is compromised and these correlations might be spurious. We next examine the correlation circle plots for both scenarios:

```
plotVar(subset1.rcc.nutrimouse)
```

```
plotVar(subset2.rcc.nutrimouse)
```

In Figure 11.3, we observe the consequences of large data sets in CCA, as the variables appear closer to the smaller circle when the number of variables increases. Hence, CCA is unable to extract relevant information (i.e. variables that explain each component), nor the correlation between the two data types.

11.5.4   Example: rCCA

We conduct rCCA on the whole data set to identify lipids and genes that best explain the correlation between both data sets.

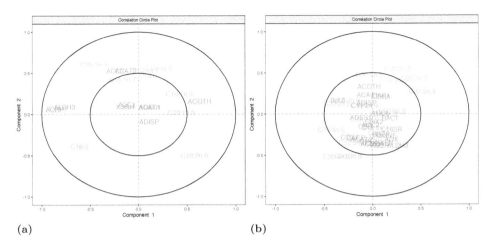

(a)   (b)

FIGURE 11.3: Classical CCA for a subset of the nutrimouse study: correlation circle plots. (a) When we consider $P = 10$ and $Q = 10$ (i.e. $N > P + Q$); (b) When we consider $P = 21$ and $Q = 25$ (i.e. $N << P + Q$). Variables tend to be closer to the large circle when $N > P + Q$ compared to when $N << P + Q$. Lipids are indicated in blue and genes in orange.

```
X <- nutrimouse$lipid
Y <- nutrimouse$gene
dim(X); dim(Y)
```

```
## [1] 40 21
```

```
## [1]   40 120
```

11.5.4.1   Cross-correlation matrix

In CCA, we consider three correlation matrices to highlight correlations between X and Y: variable correlations within X, within Y, and between both data sets $X^T Y$ (correlations between lipids and genes here).

The `imgCor()` function proposes two ways of representing the correlation structure of the data:

- Either by calculating the cross-correlations between the data concatenated as X and Y (argument `type = "combine"`);

- Or by calculating the correlation matrix of each data set separately (default argument `type = "separate"`).

Here we show the cross-correlation on the concatenated data, which is the default option (Figure 11.4):

```
imgCor(X, Y, sideColors = c('orange', 'gray'))
```

Figure 11.4 provides some insight into the correlation structure of each data set when

[X,Y] correlation matrix

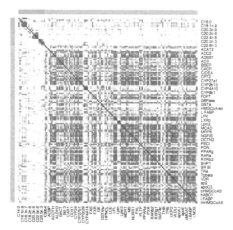

FIGURE 11.4: Cross-correlation matrix in the `nutrimouse` study between the gene and lipid data sets. The outer orange (gray) rectangles represent lipids (genes), while the inner square matrix represents the correlation coefficients, ranging from blue for negative correlations to red for positive correlations. We observe some correlation structure, mostly positive (dark red clusters) within the data set (along the diagonal) and between data sets (outside the diagonal). Note that the representation is symmetrical around the diagonal, so we only need to interpret the upper or lower side of this plot.

considered together (clusters outside the diagonal) or separately (close to the diagonal). This figure is suitable when the number of variables is not too large. We will come back to this plot later in this case study.

11.5.4.2 Regularisation parameters

Two options are proposed to choose the regularisation parameters λ_1 and λ_2 using either a cross-validation (CV) procedure (`tune.rcc()`), or the shrinkage method (directly in `rcc()`).

We first present the CV method where we define a grid of values to be tested for each parameter. Here we choose an equally spaced grid. The grid is dependent on the total number of variables in each data set (the smaller P the smaller λ_1) and may require several trials (first with a coarse grid, then by refining). The graphic outputs the first canonical correlation based on cross-validation for each possible parameter combination (Figure 11.5). Here we use leave-one-out CV:

```
grid1 <- seq(0, 0.2, length = 5)
grid2 <- seq(0.001, 0.2, length = 5)
cv.tune.rcc.nutrimouse <- tune.rcc(X, Y, grid1 = grid1,
                                   grid2 = grid2,
                                   validation = "loo")
```

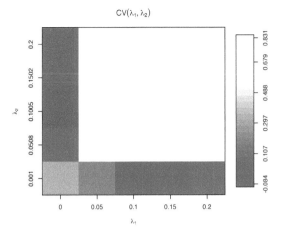

FIGURE 11.5: Choosing the regularisation parameters λ_1 and λ_2 in rCCA on the nutrimouse lipid concentration and gene expression data sets. Values in the heatmap represent the first canonical correlation on the left-out sets using CV for all possible combinations of parameter values. The highest CV score corresponding to the optimal λ_1 and λ_2 values can then be extracted as an input in rCCA.

```
cv.tune.rcc.nutrimouse
```

```
##
## Call:
##   tune.rcc(X = X, Y = Y, grid1 = grid1, grid2 = grid2, validation = "loo")
##
##    lambda1 =  0.1 , see object$opt.lambda1
##    lambda2 =  0.05075 ,  see object$opt.lambda2
##  CV-score =  0.8502 , see object$opt.score
```

Based on the CV output, we can then extract the optimal regularisation values and input them in rCCA:

```
rcc.nutrimouse <- rcc(X, Y, lambda1 = cv.tune.rcc.nutrimouse$opt.lambda1,
                   lambda2 = cv.tune.rcc.nutrimouse$opt.lambda2)
```

The next step is to choose the number of dimensions (pairs of canonical variates) to retain. The screeplot outputs all canonical correlations for an rCCA model (the total possible number of dimensions being equal to $\min(P, Q, N)$) in Figure 11.6.

```
plot(rcc.nutrimouse, type = "barplot")
```

11.5.4.3   Comparison with the shrinkage method

In comparison, this is the screeplot output from the shrinkage method (Figure 11.7):

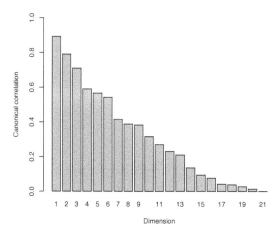

FIGURE 11.6: Screeplot of the canonical correlations when using the cross-validation method to estimate the regularisation parameters in rCCA on the `nutrimouse` lipid concentration and gene expression data sets. Using the 'elbow' rule, we could choose three dimensions.

```
rcc.nutrimouse2 <- rcc(X,Y, method = 'shrinkage')
plot(rcc.nutrimouse2, type = "barplot")
```

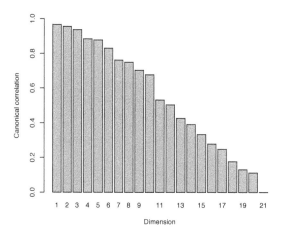

FIGURE 11.7: Screeplot of the canonical correlations when using the shrinkage method to estimate the regularisation parameters in rCCA. Compared to the cross-validation method, we observe canonical correlations close to 1 with no sharp decrease.

The plot shows that there is no sharp decrease in the canonical correlations, making it difficult to decide on the number of dimensions to choose. If we are aiming for dimension reduction, we may still choose three dimensions. We will compare next whether this approach is able to extract relevant associations between variables.

The regularised parameters are estimated as:

```
rcc.nutrimouse2$lambda
```

```
## lambda1 lambda2
##    0.139   0.136
```

11.5.4.4   Final rCCA

We choose three components for our final two models (either based on CV or the shrinkage method):

```
CV.rcc.nutrimouse <- rcc(X,Y, ncomp = 3,
                  lambda1 = cv.tune.rcc.nutrimouse$opt.lambda1,
                  lambda2 = cv.tune.rcc.nutrimouse$opt.lambda2)
shrink.rcc.nutrimouse <- rcc(X,Y, ncomp = 3, method = 'shrinkage')
```

11.5.4.5   Sample plots

Since rCCA is an unsupervised approach, the information about the treatments or groups is not taken into account in the analysis. However, we can still represent the different diets by colour, and indicate as text the genotype, to understand clusters of samples. Here the canonical variates are first averaged (argument `rep.space = "XY-variate"`) before representing the projection of the samples (Figure 11.8).

```
plotIndiv(CV.rcc.nutrimouse, comp = 1:2,
          ind.names = nutrimouse$genotype,
          group = nutrimouse$diet, rep.space = "XY-variate",
          legend = TRUE, title = 'Nutrimouse, rCCA CV XY-space')
```

```
plotIndiv(shrink.rcc.nutrimouse, comp = 1:2,
          ind.names = nutrimouse$genotype,
          group = nutrimouse$diet, rep.space = "XY-variate",
          legend = TRUE, title = 'Nutrimouse, rCCA shrinkage XY-space')
```

Projecting the samples into the XY-space shows an interesting clustering of the genotypes along the first dimension. We also observe some distinction between the five types of diet on the second dimension, more so with the CV method than the shrinkage method – the latter primarily separates the coconut diet.

The XY-space makes sense for an rCCA analysis as we aim to maximise the correlation between the components from each space. By default, the plot shows the data in their respective X- or Y-space. One way to check that both subspaces are correlated is to use the `plotArrow()` function that superimposes the two sets of canonical variates (see Section 6.2 and `?plotArrow`) as shown in Figure 11.9.

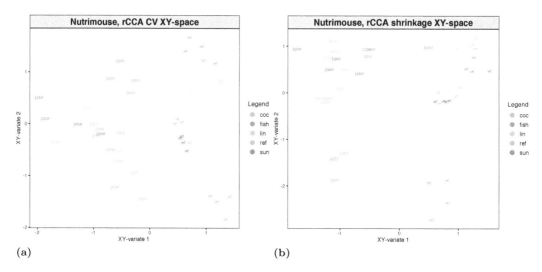

FIGURE 11.8: rCCA sample plots from the CV or shrinkage method. Canonical variates corresponding to each data set are first averaged using the argument `rep.space = "XY-variate"`. Samples are projected into the space spanned by the averaged canonical variates and coloured according to genotype information. The shrinkage method seems better at characterising the coconut diet, but not the other types of diets. A strong separation is observed between the genotypes (indicated as 'ppar' or 'wt' in text).

```
plotArrow(CV.rcc.nutrimouse, group = nutrimouse$diet,
          col.per.group = color.mixo(1:5),
          title = 'Nutrimouse, CV method')
```

```
plotArrow(shrink.rcc.nutrimouse, group = nutrimouse$diet,
          col.per.group = color.mixo(1:5),
          title = 'Nutrimouse, shrinkage method')
```

The arrow plots show the level of agreement or disagreement between the canonical variates. Interestingly, the agreement seems stronger in the shrinkage rCCA method.

3D visualisations are also offered using the following command lines (Figure not shown). Here the type of diet is represented with different colours (note that this plot requires a model run with at least three components!):

```
col.diet <- color.mixo(as.numeric(nutrimouse$diet))
plotIndiv(CV.rcc.nutri, comp = 1:3, ind.names = FALSE,
          col = color.mixo(nutrimouse$diet), style = '3d')
```

11.5.4.6   Variable plots

Here we mainly focus on the correlation circle plots and provide the command lines for complementary relevance networks and clustered image maps.

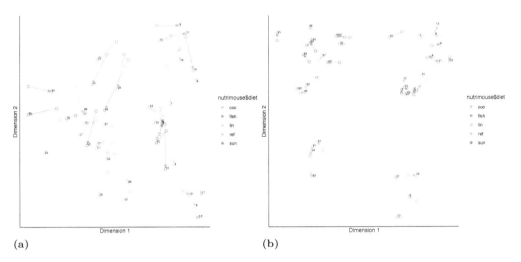

(a)   (b)

FIGURE 11.9: Arrow sample plots from the rCCA performed on the `nutrimouse` data to represent the samples projected onto the first two canonical variates. Each arrow represents one sample. The start of the arrow indicates the location of the sample in the space spanned by the variates associated to X (gene expression), and the tip the location of that same sample in the space spanned by the variates associated to Y (lipid concentration). Long arrows, as observed in the CV method rCCA in **(a)**, indicate some disagreement between the two data sets, and short arrows, as in the shrinkage method rCCA in **(b)**, strong agreement.

11.5.4.6.1 Correlation circle plots

We use correlation circle plots to visualise the correlation between variables, here genes and lipids. Both 2D and 3D plots are available (`style = '3d'` in the `plotVar()` function). Showing all variables would make it a busy plot, hence we set a cutoff based on their correlation with each canonical variate (i.e. distance from the origin). For example for a `cutoff = 0.5`, we obtain for each rCCA (Figure 11.10):

```
plotVar(CV.rcc.nutrimouse, var.names = c(TRUE, TRUE),
        cex = c(4, 4), cutoff = 0.5,
        title = 'Nutrimouse, rCCA CV comp 1 - 2')
```

```
plotVar(shrink.rcc.nutrimouse, var.names = c(TRUE, TRUE),
        cex = c(4, 4), cutoff = 0.5,
        title = 'Nutrimouse, rCCA shrinkage comp 1 - 2')
```

The correlation circle plots show that the main correlations between genes and lipids appear on the first dimension, for example: GSTPi2 with C20.5n.3 and C22.6n.3 in the top right corner; THIOL and C.16.0 in the far right and CAR1, ACOTH and C.20.1n.9 in the far left of both plots. Differences in interpretation appear in the second dimension, as we had already highlighted in the sample plots between the two methods. Thus, it will be necessary

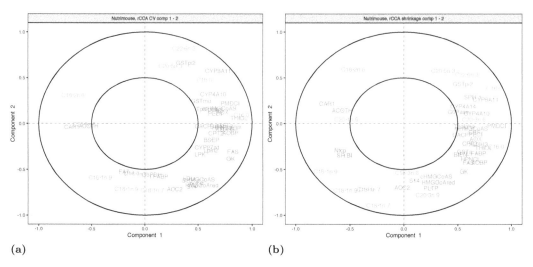

FIGURE 11.10: Correlation circle plots from the rCCA performed on the nutrimouse data showing the correlation between the gene expression and lipid concentration data. Coordinates of variables from both data sets are calculated as explained in Section 6.2. Variables with a strong association to each canonical variate are projected in the same direction from the origin and cluster. The greater the distance from the origin, the stronger the relation. To lighten the plot, only variables with a coordinate/correlation $>|0.5|$ are represented. **(a)** rCCA with the CV method and **(b)** with the shrinkage method. Both plots show a similar correlation structure on the first dimension, and differ on the second dimension, as discussed in text. Lipids and genes are coloured accordingly.

to choose the appropriate method based on biological considerations (*Is it important to separate all diets?*) and numerical considerations (*Is the high canonical correlation spurious? Does the biology make sense between the gene/lipid associations?*)

11.5.4.6.2 Relevance networks and CIM

We provide the command lines for a CIM based on rCCA with three components, which would complement the 2D correlation circle plots. If the margins are too large, consider either opening a X11() window, or saving the plot:

```
# X11() # To open a new window
cim(CV.rcc.nutrimouse, comp = 1:3, xlab = "genes", ylab = "lipids",
    # To save the plot, otherwise comment out:
    save = 'pdf', name.save = 'CIM_nutrimouse'
    )
```

We can also use relevance networks to focus on the relationship between the two types of variables (see Section 6.2).

The relevance network is built on the same similarity matrix as used by the CIM. An interactive option is available to change the cutoff (works with Windows or Linux environments).

```
# X11() # To open a new window
network(CV.rcc.nutrimouse, comp = 1:3, interactive = FALSE,
        lwd.edge = 2,
        # To save the plot, otherwise comment out:
        save = 'pdf', name.save = 'network_nutrimouse')
```

The network can be saved in a .glm format to be input into the software Cytoscape, using the R package `igraph` (Csardi et al., 2006):

```
library(igraph)
relev.net <- network(CV.rcc.nutrimouse, comp = 1:3, interactive = FALSE,
                     threshold = 0.5)
write.graph(relev.net$gR, file = "network.gml", format = "gml")
```

11.5.4.6.3 Back to the cross-correlation matrix

Previously, we examined the cross-correlation between both full data sets. We can extract the name of the lipids and genes that appear correlated on the `plotVar()` plot in each data set, then output the image plot `imgCor()`:

```
# Extract coordinates by saving plotVar into an object
var.coord <- plotVar(CV.rcc.nutrimouse, var.names = c(TRUE, TRUE),
        cutoff = 0.6, plot = FALSE)

list.lipid <- var.coord$names[var.coord$Block == 'X']
list.gene <- var.coord$names[var.coord$Block == 'Y']

imgCor(X[, list.lipid], Y[, list.gene], sideColors = c('orange', 'gray'))
```

Figure 11.11 highlights a better correlation now that we have been able to identify the key correlated molecules.

11.6 To go further

As we mentioned previously in Section 10.7, most sparse CCA approaches have been solved using penalised PLS decomposition. We refer the reader to the R packages `PMA` (Witten and Tibshirani, 2020) (that uses a PLS-based algorithm) and `sRDA` (Csala, 2017) in particular, and the FAQ below comparing rCCA with PLS canonical mode. Solving sparse CCA with other formulations is a difficult problem. Other approaches have been proposed, such as those from Hardoon and Shawe-Taylor (2011), Wilms and Croux (2015) and Wilms and Croux (2016), but those approaches are not implemented into software.

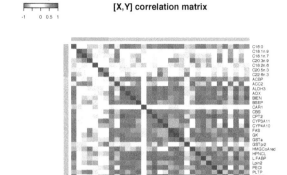

FIGURE 11.11: Cross-correlation matrix only on the lipids and genes that most contribute to each canonical variate. The correlation structure is better highlighted compared to Figure 11.4 that included all genes and lipids.

11.7   FAQ

Can CCA and rCCA manage missing values?

Yes, for classical CCA and rCCA where we input the CV-tuned regularised parameters. No, for rCCA with the shrinkage approach, where you will need to impute missing values.

Can CCA and rCCA manage more than two data sets?

No, CCA can only manage two data sets, but we propose other alternatives in Chapter 13.

Can I perform CCA when I have more variables than samples

No, CCA is not reliable when $P + Q >> N$ because of computational issues. For large data sets, consider using rCCA.

Can (r)CCA fail?

(r)CCA may not bring relevant results when:

- There is no correlated information between the two data sets.
- Linear combinations of both types of data cannot unravel a common correlation structure between the two data sets.
- The shrinkage parameters in rCCA are not well chosen.

What about repeated measurements? The multilevel data decomposition described in Section 4.1 can be applied here externally, using the `withinVariation()` function to first extract the within variance data that is then input into CCA or rCCA.

What about compositional data? Data can be transformed with Centered Log Ratio using the `logratio.transfo()` function first (you may need to specify an `offset` value if your data contain zeros), as described in Section 4.1. The CLR data can then be input in CCA or rCCA.

Besides computational aspects, what is the difference between a PLS canonical mode (Section 10.2) and rCCA? The underlying rCCA algorithm is vastly different from the PLS algorithm (see the pseudo CCA algorithm in Appendix 11.A). In terms of data analysis and interpretation, components will also differ. PLS focuses on the common sources of variation (in **nutrimouse** the genotype effect). Even when centering and scaling the data in PLS, both methods may bring different results. In **nutrimouse**, components are highly correlated between both approaches when we consider the shrinkage rCCA, but not the CV rCCA. The correlation circle plots seem to highlight the same strongly correlated variables. One advantage of PLS is the variable selection process. We append below some of the code to compare both approaches (outputs not shown).

```
# PLS canonical mode
pls.nutrimouse <- pls(X, Y, ncomp = 3, mode = 'canonical')

# Comparison of the sample plots between PLS and rCCA with shrinkage
plotIndiv(pls.nutrimouse, group = nutrimouse$genotype,
          ind.names = nutrimouse$diet)
plotIndiv(shrink.rcc.nutrimouse, group = nutrimouse$genotype,
          ind.names = nutrimouse$diet)

# Correlation between canonical variates and PLS components (on diagonal)
# Corresponding to each data set
cor(pls.nutrimouse$variates$X, shrink.rcc.nutrimouse$variates$X)
cor(pls.nutrimouse$variates$Y, shrink.rcc.nutrimouse$variates$Y)

 # Correlation circle plots highlight similar groups of strongly correlated variables
plotVar(pls.nutrimouse, cutoff = 0.5)
plotVar(shrink.rcc.nutrimouse, cutoff = 0.5)
```

11.8   Summary

CCA is an approach that integrates two data sets of quantitative variables measured on the same individuals or samples where both data sets play a symmetrical role. The difference between PLS and CCA lies in the algorithm used: CCA requires the inversion of variance-covariance matrices and faces computational instability when the number of variables becomes large. In that case, a regularised version of CCA can be used, where the parameters are chosen according to either a cross-validation method, or a shrinkage approach. The other difference between CCA and PLS is that CCA maximises the correlation between canonical variates (components), whilst PLS maximises the covariance.

The approach is unsupervised and exploratory, and most CCA results are interpreted via graphical outputs. Correlation circle plots, clustered image maps and relevance networks help to identify subsets of correlated variables from both data sets but CCA does not enable variable selection.

11.A   Appendix: CCA and variants

Notations for CCA

X $(N \times P)$ and Y $(N \times Q)$ are two data matrices matched on the same samples N. We denote by X^j and Y^k the vector variables in the X and the Y data sets ($j = 1, \ldots, P$ and $k = 1, \ldots, Q$). In classical CCA, we assume that $P \leq Q$ (the data set that contains the fewest variables is denoted X). We denote S_{XX} and S_{YY} of respective size $(P \times P)$ and $(Q \times Q)$ the sample variance-covariance matrices (defined in 3.1) for variables in the data sets X and Y, and by $S_{XY} = S_{YX}^T$ the $(P \times Q)$ sample cross-covariance matrix between X and Y. We denote I_P the identity matrix, i.e. a $(P \times P)$ matrix with ones along the diagonal and zeros for all other entries.

11.A.1   Solving classical CCA

Classical CCA assumes that $P \leq N$, $Q \leq N$ and that the matrices X and Y are of full column rank P and Q, respectively. In the following, the principle of CCA is presented as a problem solved through an iterative algorithm, but we will see later that it is solved into one eigenvalue problem.

The first stage of CCA consists of finding two vectors $a^1 = (a_1^1, \ldots, a_P^1)'$ and $b^1 = (b_1^1, \ldots, b_Q^1)'$ that maximises the correlation between the linear combinations

$$t^1 = Xa^1 = a_1^1 X^1 + a_2^1 X^2 + \cdots + a_P^1 X^P$$

and

$$u^1 = Yb^1 = b_1^1 Y^1 + b_2^1 Y^2 + \cdots + b_q^1 Y^Q$$

assuming that the vectors a^1 and b^1 are normalised so that the variance of the canonical variates is 1, i.e. $\operatorname{var}(t^1) = \operatorname{var}(u^1) = 1$.

The problem consists in solving:

$$\rho_1 = \operatorname{cor}(t^1, u^1) = \underset{a,b}{\operatorname{argmax}} \operatorname{cor}(Xa, Yb),$$

subject to the constraints $\operatorname{var}(Xa) = \operatorname{var}(Yb) = 1$. The resulting variables t^1 and u^1 are called the first *canonical variates* and ρ_1 is the first canonical correlation.

Higher order canonical variates and canonical correlations can be found as a stepwise problem. For $h = 1, \ldots, P$, we can find positive correlations $\rho_1 \geq \rho_2 \geq \cdots \geq \rho_P$ with corresponding pairs of vectors $(a^1, b^1), \ldots, (a^P, b^P)$, successively by maximizing:

$$\rho_h = \operatorname{cor}(t^h, u^h) = \underset{a^h, b^h}{\operatorname{argmax}} \operatorname{cor}(Xa^h, Yb^h) \quad \text{subject to} \quad \operatorname{var}(Xa^h) = \operatorname{var}(Yb^h) = 1,$$

under the additional restriction that $\text{cor}(\boldsymbol{t}^s, \boldsymbol{t}^h) = \text{cor}(\boldsymbol{u}^s, \boldsymbol{u}^h) = 0$ for $1 \leq h < s \leq P$ (i.e. the canonical variates associated to each data set are uncorrelated).

Note: Contrary to what we have seen in PLS, there is no deflation of the data matrices.

11.A.1.1 Geometrical point of view

We define the orthogonal projectors onto the linear combinations of the $\boldsymbol{X}$ and $\boldsymbol{Y}$ variables as:

$$P_X = \boldsymbol{X}(\boldsymbol{X}^T\boldsymbol{X})^{-1}\boldsymbol{X}^T = \frac{1}{N}\boldsymbol{X}\boldsymbol{S}_{XX}^{-1}\boldsymbol{X}^T$$

and

$$P_Y = \boldsymbol{Y}(\boldsymbol{Y}^T\boldsymbol{Y})^{-1}\boldsymbol{Y}^T = \frac{1}{N}\boldsymbol{Y}\boldsymbol{S}_{YY}^{-1}\boldsymbol{Y}^T$$

We have the following properties (Mardia et al., 1979):

- The canonical correlations ρ_h are the positive square roots of the eigenvalues λ_s of $P_X P_Y$ (which are the same as those of $P_Y P_X$): $\rho_h = \sqrt{\lambda_h}$,
- The vectors $\boldsymbol{t}^1, \ldots, \boldsymbol{t}^P$ are the standardised eigenvectors corresponding to the decreasing eigenvalues $\lambda_1 \geq \cdots \geq \lambda_P$ of $P_X P_Y$; and,
- The vectors $\boldsymbol{u}^1, \ldots, \boldsymbol{u}^P$ are the standardised eigenvectors corresponding to the same decreasing eigenvalues of $P_Y P_X$.

11.A.1.2 Pseudo algorithm

The classical CCA solution to the eigenvalues problem is (Hotelling, 1936):

$$\begin{aligned}
\boldsymbol{S}_{XX}^{-1}\boldsymbol{S}_{XY}\boldsymbol{S}_{YY}^{-1}\boldsymbol{S}_{YX}\boldsymbol{a} &= \lambda^2 \boldsymbol{a}, \\
\boldsymbol{S}_{YY}^{-1}\boldsymbol{S}_{YX}\boldsymbol{S}_{XX}^{-1}\boldsymbol{S}_{XY}\boldsymbol{b} &= \lambda^2 \boldsymbol{b}.
\end{aligned}$$

In practice, we calculate the canonical variates and canonical correlations as follows:

CALCULATE the upper triangular matrices $\boldsymbol{L}_X$ and $\boldsymbol{L}_Y$ verifying
$\boldsymbol{S}_{XX} = \boldsymbol{L}_X \boldsymbol{L}_X^T, \quad \boldsymbol{S}_{YY} = \boldsymbol{L}_Y \boldsymbol{L}_Y^T$ using Cholesky decomposition[1].
CALCULATE $\boldsymbol{M} = \boldsymbol{L}_Y^{-1} \boldsymbol{S}_{YX} (\boldsymbol{L}_X^T)^{-1}$.
CALCULATE the singular value decomposition of $\boldsymbol{M}$: $\boldsymbol{M} = \boldsymbol{T}_M \boldsymbol{D}_M \boldsymbol{U}_M^T$.
The canonical correlations are on the diagonal of $\boldsymbol{D}_M$
The canonical variates are calculated as: $\boldsymbol{T} = \boldsymbol{X}(\boldsymbol{L}_X^T)^{-1}\boldsymbol{U}_M$ and $\boldsymbol{U} = \boldsymbol{Y}(\boldsymbol{L}_Y^T)^{-1}\boldsymbol{T}_M$.

11.A.2 Regularised CCA

In classical CCA, calculating the canonical correlations and canonical variates requires the inversion of $\boldsymbol{S}_{XX}$ and $\boldsymbol{S}_{YY}$. When $P >> N$, the sample variance-covariance matrix $\boldsymbol{S}_{XX}$ is singular, and similarly for $\boldsymbol{S}_{YY}$ when $Q >> N$.

---

[1]A decomposition into a lower triangular matrix and its conjugate transpose with positive diagonal entries.

In addition, even when $P << N$, variables might be highly correlated within $\boldsymbol{X}$, resulting in $\boldsymbol{S}_{XX}$ to be ill-conditioned. The calculation of its inverse is unreliable - likewise for $\boldsymbol{S}_{YY}$ (Friedman, 1989). This phenomenon also occurs when the ratio $\frac{P}{N}$ (or $\frac{Q}{N}$) is less than one but not negligible. Thus, a standard condition for CCA is that $N > P + Q$. If we do not meet this assumption, we must consider regularised CCA that uses ridge-type regularisation.

11.A.2.1   General principle of rCCA

The principle of ridge regression (Hoerl and Kennard, 1970) was extended to CCA by Vinod (1976), then by Leurgans et al. (1993). It involves the regularisation of the variance-covariance matrices by adding a small value on their diagonal to make them invertible:

$$\boldsymbol{S}_{XX}(\lambda_1) = \boldsymbol{S}_{XX} + \lambda_1 \boldsymbol{I}_P$$

and

$$\boldsymbol{S}_{YY}(\lambda_2) = \boldsymbol{S}_{YY} + \lambda_2 \boldsymbol{I}_Q,$$

where λ_1 and λ_2 are non-negative numbers such that $\boldsymbol{S}_{XX}(\lambda_1)$ and $\boldsymbol{S}_{YY}(\lambda_2)$ become regular (well conditioned) matrices. Regularised CCA then simply consists of performing classical CCA substituting $\boldsymbol{S}_{XX}$ and $\boldsymbol{S}_{YY}$, respectively with their regularised versions $\boldsymbol{S}_{XX}(\lambda_1)$ and $\boldsymbol{S}_{YY}(\lambda_2)$. Of course this step requires the tuning of the regularisation parameters λ_1 and λ_2.

11.A.2.2   Tuning regularisation parameters with the cross-validation method

González et al. (2008) proposed to extend the leave-one-out procedure suggested by Leurgans et al. (1993) and to choose the regularisation parameters λ_1 and λ_2 with M-fold cross-validation. We have already covered similar strategies to tune the lasso parameters in sPCA (see Section 9.B and sPLS (Section 10.C). Here we aim to choose the best set of parameters to maximise the first canonical correlation (the correlation between canonical variates) on the left-out set.

The process is as follows:

CHOOSE a grid of values for λ_1 and λ_2
    FOR EACH M-fold using cross-validation
        Divide the N samples into training and testing sets used to obtain
            $\boldsymbol{X}_{\text{train}}, \boldsymbol{Y}_{\text{train}}, \boldsymbol{X}_{\text{test}}, \boldsymbol{Y}_{\text{test}}$
        FOR EACH combination of parameters (λ_1, λ_2)
            Calculate the regularised variance-covariance matrices $\boldsymbol{S}_{XX}(\lambda_1)$ and $\boldsymbol{S}_{YY}(\lambda_2)$ on the training sets
            Solve rCCA on $\boldsymbol{X}_{\text{train}}$ and $\boldsymbol{Y}_{\text{train}}$ using $\boldsymbol{S}_{XX}(\lambda_1)$ and $\boldsymbol{S}_{YY}(\lambda_2)$
            Extract the first canonical factors $\boldsymbol{a}^1_{\text{train}}$ and $\boldsymbol{b}^1_{\text{train}}$ associated to $\boldsymbol{X}_{\text{train}}$ and $\boldsymbol{Y}_{\text{train}}$
            Calculate the predicted canonical variates $\boldsymbol{t}^1_{\text{test}} = \boldsymbol{X}_{\text{test}} \boldsymbol{a}^1_{\text{train}}$ and $\boldsymbol{u}^1_{\text{test}} \leftrightarrow \boldsymbol{Y}_{\text{test}} \boldsymbol{b}^1_{\text{train}}$
            Calculate the predicted canonical correlation $\rho_{(\lambda_1,\lambda_2)} = \text{cor}(\boldsymbol{t}^1_{\text{test}}, \boldsymbol{u}^1_{\text{test}})$
        Average the predicted canonical correlations $\bar{\rho}_{(\lambda_1,\lambda_2)}$ across the M folds
        Choose (λ_1, λ_2) for which $\bar{\rho}_{(\lambda_1,\lambda_2)}$ is maximised

Notes:

- *In practice, we choose $M = 3 : 10$, depending on N.*

- (λ_1, λ_2) *are chosen with respect to the first canonical variates and are then fixed for higher order canonical variates. When N is very small (<10), we can use leave-one-out cross-validation, however, the computational burden can be considerable on some data sets. To alleviate computational time, the shrinkage method can be used.*

11.A.2.3 Estimating the regularisation parameters with the shrinkage method

Using the method of Schäfer and Strimmer (2005) implemented in the `corpcor` package (Schafer et al., 2017), we compute the empirical variance of each variable in $\boldsymbol{X}$ and shrink the variances towards their median. The corresponding shrunk parameter λ_1 (for the $\boldsymbol{X}$ matrix) is estimated as:

$$\lambda_1 = \frac{\sum_{j=1}^{P} \mathrm{var}(\boldsymbol{X}^j)}{\sum_{j=1}^{P} (\mathrm{var}(\boldsymbol{X}^j) - \mathrm{median}(\mathrm{var}(\boldsymbol{X}^j)))^2}$$

We do similarly for the $\boldsymbol{Y}$ matrix to estimate λ_2. The approach is fast to compute and should be appropriate for small data sets, according to the authors.

Note:

- *This approach estimates the regularisation parameters on each data set individually.*

12

PLS-Discriminant Analysis (PLS-DA)

12.1 Why use PLS-DA?

PLS-Discriminant Analysis (PLS-DA) is a linear multivariate model which performs classification tasks and is able to predict the class of new samples. As the name suggests, the method extends PLS from integrating two continuous data matrices to integrating a continuous data matrix X with a categorical outcome variable. PLS-DA seeks for components that best separate the sample groups, whilst the sparse version also selects variables that best discriminate between groups. This Chapter illustrates these two approaches on the srbct tumour gene expression study, gives examples for sample class prediction and how to apply these methods for repeated measurement designs or microbiome data. We also discuss how to minimise the effect of overfitting in this context.

12.1.1 Biological question

Can I discriminate samples based on their outcome category? Which variables discriminate the different outcomes? Can they constitute a molecular signature that predicts the class of external samples?

12.1.2 Statistical point of view

These questions suggest a supervised framework, where the response is categorical and known. Therefore, the input in PLS-DA should be a *matrix* of numerical variables (X) and a *categorical vector* indicating the class of each sample (denoted y). We consider two types of approaches:

- PLS-DA, which is a special case of PLS where the outcome vector y is then transformed into a matrix of indicator variables assigning each sample to a known class,

- sparse PLS-DA (Lê Cao et al., 2011), which uses lasso penalisation similar to sPLS to select variables in the X data matrix (refer to Section 10.2).

In this chapter, we use cross-validation to evaluate the performance of the model and tune the parameters.

TABLE 12.1: Example of an outcome factor (left) transformed into an indicator dummy matrix (right). The transformation is performed internally in our supervised methods using the unmap() function in the package. Note that the third column is actually redundant as it can be built from the first two.

Samples	Treatment	Samples	trt1	trt2	trt3
indiv_1	trt_1	indiv_1	1	0	0
indiv_2	trt_2	indiv_2	0	1	0
indiv_3	trt_3	indiv_3	0	0	1
indiv_4	trt_1	indiv_4	1	0	0
indiv_5	trt_2	indiv_5	0	1	0
indiv_6	trt_3	indiv_6	0	0	1
indiv_7	trt_1	indiv_7	1	0	0
indiv_8	trt_2	indiv_8	0	1	0
indiv_9	trt_3	indiv_9	0	0	1
indiv_10	trt_1	indiv_10	1	0	0

12.2 Principle

Projection to Latent Structures (PLS) was principally designed for regression problems, where the outcome, response Y, is continuous. However, PLS is also known to perform very well for classification problems where the outcome is categorical (Ståhle and Wold, 1987). Historically, Fisher's Discriminant Analysis (also called Linear Discriminant Analysis, LDA) was commonly used for classification problems, but this method is limited for large data sets as it requires computing the inverse of a large variance-covariance matrix. As we have seen previously with PLS in Section 10.2, PLS-DA circumvents this issue with the use of latent components and local regressions. PLS-DA can handle multi-class problems (more than two sample groups) without having to decompose the problem into several two-class subproblems[1].

To incorporate a categorical vector into a PLS-type regression model, we first need to transform the vector y into a *dummy* matrix denoted Y that records the class membership of each sample: each response category is coded via an indicator variable, as shown in Table 12.1, resulting in a matrix with N rows and K columns corresponding to the K sample classes. The PLS regression (now PLS-DA) is then run with the outcome treated as a continuous matrix. This trick works well in practice, as demonstrated in several references (Nguyen and Rocke (2002b), Tan et al. (2004), Nguyen and Rocke (2002a), Boulesteix and Strimmer (2007), Chung and Keles (2010)).

We use the following notations: X is an $(N \times P)$ data matrix, y is a factor vector of length N that indicates the class of each sample, and Y is the associated dummy $(N \times K)$ data matrix, with N the number of samples, P the number of predictor variables and K the number of classes.

Similar to PLS, PLS-DA outputs the following (also see Figure 12.1):

---

[1]The term PLS-DA originates from Ståhle and Wold (1987), Indahl et al. (2007) with Indahl et al. (2009) who considered PLS-DA as a special case of Canonical PLS.

- A set of components, or latent variables. There are as many components as the chosen dimensions of the PLS-DA model.
- A set of loading vectors, which are coefficients corresponding to each variable. The coefficients are defined so that linear combinations of variables maximise the discrimination between groups of samples. To each PLS-DA component corresponds a loading vector.

Since we use a PLS algorithm, we also obtain components and loading vectors associated to the Y dummy matrix. However, those are not of primary interest for the classification task at hand.

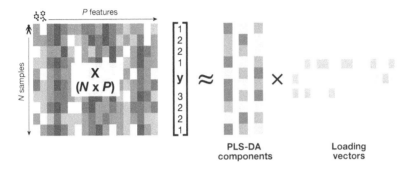

FIGURE 12.1: Schematic view of the PLS-DA matrix decomposition into sets of components (latent variables) and loading vectors. y is the outcome factor that is internally coded as a dummy matrix Y in the function, as shown in Table 12.1. In this diagram, only the components and loading vectors associated to the X predictor matrix are shown, and are of interest.

12.2.1 PLS-DA

12.2.1.1 PLS regression mode with a dummy matrix Y

Similar to PLS, PLS-DA maximises the covariance between a linear combination of the variables from X and a linear combination of the dummy variables from Y. For example, for the first dimension, the objective function to solve is:

$$\underset{||a||=1,||b||=1}{\mathrm{argmax}} \quad \mathrm{cov}(Xa, Yb), \tag{12.1}$$

where a and b are the loading vectors and $t = Xa$ and $u = Yb$ are their corresponding latent variables. In practice, we will mainly focus on the vectors a and t that correspond to the X predictor matrix. The objective function is solved with SVD, as we have described in Appendix 10.A.

Note:

- *The PLS mode is set to* `mode = regression` *internally in the function.*

12.2.1.2 Prediction of the class of a new sample

In a classification context, one of our aims is to predict the class of new samples from a trained model – either by using cross-validation or when we have access to a real test data set, as we covered in Section 7.2.

We can formulate a PLS-DA model in a matrix form as:

$$\boldsymbol{Y} = \boldsymbol{X}\boldsymbol{\beta} + \boldsymbol{E}$$

where $\boldsymbol{\beta}$ is the matrix of regression coefficients and $\boldsymbol{E}$ is a residual matrix. The prediction of a new set of samples is:

$$\boldsymbol{Y}_{\mathrm{new}} = \boldsymbol{X}_{\mathrm{new}}\boldsymbol{\beta}$$

Because the prediction is made on the dummy variables from $\boldsymbol{Y}_{\mathrm{new}}$, a prediction distance is needed to map back to a $\boldsymbol{Y}$ outcome, as we introduced in Section 7.5. Several distances are implemented in `mixOmics` to predict the class membership of each new sample (each row in $\boldsymbol{Y}_{\mathrm{new}}$). For example, we can assign the predicted class as the column index of the element with the largest predicted value in this row (`method.predict = 'max.dist'` in the `predict()` function). A detailed example is given in Section 12.5, see also Appendix 12.A.

12.2.2   sparse PLS-DA

The sparse variant sPLS-DA enables the selection of the most predictive or discriminative features in the data to classify the samples (Lê Cao et al., 2011). sPLS-DA performs variable selection and classification in a one-step procedure. It is a special case of sparse PLS, where the lasso penalisation applies only on the loading vector $\boldsymbol{a}$ associated to the $\boldsymbol{X}$ data set.

Let $\boldsymbol{M} = \boldsymbol{X}^T\boldsymbol{Y}$, for the first dimension, we solve

$$\underset{||\boldsymbol{a}||=1}{\mathrm{argmin}} \parallel \boldsymbol{M} - \boldsymbol{a}\boldsymbol{b}^T \parallel_F^2 + P_\lambda(\boldsymbol{a}), \qquad (12.2)$$

where P_λ is a lasso penalty function with regularisation parameter λ, and $\boldsymbol{a}$ is the sparse loading vector that enables variable selection. For the following dimensions, the $\boldsymbol{X}$ and $\boldsymbol{Y}$ matrices are deflated as in PLS (see Appendix 10.A).

12.3   Input arguments and tuning

12.3.1   PLS-DA

For PLS-DA, we need to specify the number of dimensions or components (argument `ncomp`) to retain. Typically, each dimension tends to focus on the discrimination of one class vs. the others, hence, if $\boldsymbol{y}$ includes K classes, `ncomp = K-1` might be sufficient. However, it depends on the difficulty of the classification task. The `perf()` function uses repeated cross-validation and returns the classification performance (classification error rate, see Section 7.3) for each component. The output of this function will also help in choosing the appropriate prediction distance that gives the best performance (and will be used to tune the parameters in sPLS-DA). We also recommend using the Balanced Error Rate described in Section 7.3 in the case where samples are not represented in equal proportions across all classes.

12.3.2 sPLS-DA

Once we have chosen the number of components and the prediction distance, we need to specify the number of variables to select on each component (argument `keepX`). The `tune()` function allows us to set a grid of parameter values `keepX` to test on each component using repeated cross-validation and returns the optimal number of variables to select for each dimension, as we detailed in Section 7.3.

Note:

- *Tuning the number of variables to select cannot be accurately estimated if the number of samples is small as we use cross-validation. When $N \leq 10$ it is best to set* `keepX` *to an arbitrary choice.*

12.3.3 Framework to manage overfitting

The issue of selection bias and overfitting in variable selection problems needs to be carefully considered with PLS-DA, and sPLS-DA (Ambroise and McLachlan (2002), Ruiz-Perez et al. (2020), see also Section 1.4). Indeed, when given a large number of variables, any discriminant model can manage to weight the variables accordingly to successfully discriminate the classes. Additionally, in sPLS-DA the variables selected may not generalise to a new data set, or to the same learning data set where some samples are removed. In other words, the sPLS-DA classifier and its variable selection 'sticks' too much to the learning set. This is why we use cross-validation during the tuning process.

We propose the following framework in `mixOmics` to avoid overfitting during tuning and model fitting (see Section 12.5):

1. **Choose the number of components**

- Tune `ncomp` on a sufficient number of components using a PLS-DA model with the `perf()` function. Set the `nrepeat` argument for repeated cross-validation to ensure accurate performance estimation.
- Choose the optimal `ncomp = H` along with its prediction distance that yields, on average, the lowest misclassification error rate based on both mean and standard deviation.

2. **Choose the number of variables to select**

Now that `ncomp` is chosen, along with the optimal prediction distance, run sPLS-DA on a grid of `keepX` values (this may take some time to compute depending on the size of the grid) with the `tune()` function which internally performs the following steps for each component h:

- Performance assessment of sPLS-DA for each `keepX` value,

- Internal choice of the `keepX` value with the lowest average classification error rate, which is retained to evaluate the performance of the next component $h + 1$.

3. **Run the final model**

Run sPLS-DA with the `ncomp` and `keepX` values that are output from the steps above, or that are arbitrarily chosen by the user.

4. **Final model assessment**

Run the `perf()` function on the final sPLS-DA model with the argument `nrepeat` to assess the performance and examine the stability of the variables selected.

5. **Variable signature**

Retrieve the list of selected variables using `selectVar()`, which we can then cross-compare with the stability of those variables.

6. **Prediction (see Section 12.5)**

If an external test set is available, predict the class of the external observations using the function `predict()`.

Notes:

- *In mixOmics we use stratified CV to ensure there is approximately the same proportion of samples in each class in each of the folds.*
- *In the tune() function, the optimal keepX value that achieves the lowest error rate per component is determined via one-sided $t-$tests between one keepX value to the next, to assess whether the classification error rate significantly decreases across repeated folds.*
- *The CV procedure we use can be biased or over-optimistic (Westerhuis et al., 2008). One way to circumvent this problem is to use cross model validation ('2CV'), which includes a second stratum of cross-validation within the training set: there is an 'outer loop' with M-fold CV and for each step of the outer loop, a complete 'inner loop' with its own M-fold CV procedure (Smit et al., 2007). As this procedure is not suitable when $N \leq 50$, which is common in omics studies, we have not implemented this option in mixOmics. As a rule of thumb, we recommend using LOO-CV when $N \leq 10$.*
- *As highlighted by Ruiz-Perez et al. (2020), PLS-DA has a high propensity to overfit. CV-based permutation tests can be performed to conclude whether there is a significant difference between groups (Westerhuis et al., 2008). This is implemented by the MVA.test() function in the RVAideMemoire package, and uses 2CV (see the detailed tutorials and examples in Hervé et al. (2018)).*

12.4 Key outputs

PLS-DA and sPLS-DA generate several outputs that are rich in information and are listed below.

12.4.1   Numerical outputs

We take advantage of the supervised analysis (and the use of cross-validation to predict the class of samples) to obtain a series of insightful outputs:

- The classification performance of the final model (using `perf()`),
- The list of variables selected, and their stability (using `selectVar()` and `perf()`),
- The predicted class of new samples (detailed in Section 12.5),
- The AUC ROC which *complements* the classification performance outputs (see Section 12.5),
- The amount of explained variance per component (but bear in mind that PLS-DA does not aim to maximise the variance explained, but rather, the covariance with the outcome),
- The VIP measure for the X (selected) variables, described in Section 10.4.

12.4.2   Graphical outputs

12.4.2.1   Sample plots

1. Sample plots with `plotIndiv()` can also be overlaid with:

- Prediction backgrounds (see Section 12.5 and Appendix 12.A),
- Predicted coordinates for test samples (requires `style = 'graphics'` in `plotIndiv()`).

2. Arrow plots with `plotArrow()`, where each arrow representing a sample starts from the class centroid of all samples.

12.4.2.2   Variable plots

1. Correlation circle plots with `plotVar()`, which enable us to visualise the correlations between variables.
2. Clustered Image Maps with `cim()`, which are simple hierarchical clustering heatmaps with samples in columns and selected variables in rows, where we use a specified distance and clustering method (`dist.method` and `clust.method` arguments).
3. Relevance networks with `network()`, that link selected variables from X to each dummy variable from Y (i.e. each class).
4. Loading plots with `plotLoadings()` that output the coefficient weight of each selected variable, ranked from most important (bottom) to least important (top), as detailed in Section 6.2. Colours are assigned to each barplot to indicate which class (according to the mean, or the median `method = 'mean'` or `'median'`) is minimised or maximised (`contrib = 'min'` or `'max'`) in each variable. If saved as an object, the function will also output which class is represented for each variable.

12.5   Case study: SRBCT

The Small Round Blue Cell Tumours (SRBCT) data set from (Khan et al., 2001) includes the expression levels of 2,308 genes measured on 63 samples. The samples are divided into four classes: 8 Burkitt Lymphoma (BL), 23 Ewing Sarcoma (EWS), 12 neuroblastoma (NB), and 20 rhabdomyosarcoma (RMS). The data are directly available in a processed and normalised format from the mixOmics package and contains the following:

- $gene: A data frame with 63 rows and 2,308 columns. These are the expression levels of 2,308 genes in 63 subjects,

- $class: A vector containing the class of tumour for each individual (four classes in total),

- $gene.name: A data frame with 2,308 rows and 2 columns containing further information on the genes.

More details can be found in ?srbct. We will illustrate PLS-DA and sPLS-DA which are suited for large biological data sets where the aim is to identify molecular signatures, as well as classify samples. We will analyse the gene expression levels of srbct$gene to discover which genes may best discriminate the four groups of tumours.

12.5.1   Load the data

We first load the data from the package (see Section 8.4 to load your own data). We then set up the data so that X is the gene expression matrix and y is the factor indicating sample class membership. y will be transformed into a dummy matrix Y inside the function. We also check that the dimensions are correct and match both X and y:

```
library(mixOmics)
data(srbct)
X <- srbct$gene

# Outcome y that will be internally coded as dummy:
Y <- srbct$class
dim(X); length(Y)
```

```
## [1]   63 2308
## [1] 63
```

```
summary(Y)
```

```
## EWS  BL  NB RMS
##  23   8  12  20
```

12.5.2   Quick start

12.5.2.1   PLS-DA

```
result.plsda.srbct <- plsda(X, Y)      # 1 Run the method
plotIndiv(result.plsda.srbct)          # 2 Plot the samples
plotVar(result.plsda.srbct)            # 3 Plot the variables
```

We run `?plsda` to identify the default arguments in this function:

- `ncomp = 2`: The first two PLS-DA components are calculated,
- `scale = TRUE`: Each data set is scaled (each variable has a variance of 1 to enable easier comparison) - data are internally centered.

By default, the sample plot `plotIndiv()` will automatically colour the samples according to their class membership (retrieved internally from the input Y).

The correlation circle plot `plotVar()` appears very cluttered, hence there will be some benefit in using sPLS-DA.

12.5.2.2 sPLS-DA

Here, we specify an arbitrary number of variables to select on each dimension, namely 50 and 30.

```
splsda.result <- splsda(X, Y, keepX = c(50,30)) # 1 Run the method
plotIndiv(splsda.result)                        # 2 Plot the samples
plotVar(splsda.result)                          # 3 Plot the variables
selectVar(splsda.result, comp = 1)$name         # Selected variables on comp 1
plotLoadings(splsda.result, method = 'mean', contrib = 'max')
```

The default parameters in `?splsda` are similar to PLS-DA, with the addition of:

- `keepX`: set to P (i.e. select all variables) if the argument is not specified.

The selected variables are ranked by absolute coefficient weight from each loading vector in `selectVar()`, and the class they describe (according to the mean/median min or max expression values) is represented in `plotLoadings()`.

12.5.3 Example: PLS-DA

12.5.3.1 Initial exploration with PCA

As covered in Chapter 9, PCA is a useful tool to explore the gene expression data and to assess for sample similarities between tumour types. Remember that PCA is an unsupervised approach, but we can colour the samples by their class to assist in interpreting the PCA (Figure 12.2). Here we center (default argument) and scale the data:

```
pca.srbct <- pca(X, ncomp = 3, scale = TRUE)

plotIndiv(pca.srbct, group = srbct$class, ind.names = FALSE,
          legend = TRUE,
          title = 'SRBCT, PCA comp 1 - 2')
```

We observe almost no separation between the different tumour types in the PCA sample plot, with perhaps the exception of the NB samples that tend to cluster with other samples. This preliminary exploration teaches us two important findings:

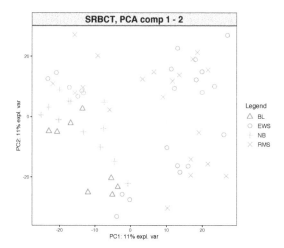

FIGURE 12.2: Preliminary (unsupervised) analysis with PCA on the SRBCT gene expression data. Samples are projected into the space spanned by the principal components 1 and 2. The tumour types are not clustered, meaning that the major source of variation cannot be explained by tumour types. Instead, samples seem to cluster according to an unknown source of variation.

- The major source of variation is not attributable to tumour type, but an unknown source (we tend to observe clusters of samples but those are not explained by tumour type).
- We need a more 'directed' (supervised) analysis to separate the tumour types, and we should expect that the amount of variance explained by the dimensions in PLS-DA analysis will be small.

12.5.3.2 Number of components in PLS-DA

The `perf()` function evaluates the performance of PLS-DA – i.e. its ability to rightly classify 'new' samples into their tumour category using repeated cross-validation. We initially choose a large number of components (here `ncomp = 10`) and assess the model as we gradually increase the number of components. Here we use three-fold CV repeated ten times. In Section 7.2 we provide further guidelines on how to choose the `folds` and `nrepeat` parameters:

```
plsda.srbct <- plsda(X,Y, ncomp = 10)

set.seed(30) # For reproducibility with this handbook, remove otherwise
perf.plsda.srbct <- perf(plsda.srbct, validation = 'Mfold', folds = 3,
                 progressBar = FALSE,  # Set to TRUE to track progress
                 nrepeat = 10)          # We suggest nrepeat = 50

plot(perf.plsda.srbct, sd = TRUE, legend.position = 'horizontal')
```

From the classification performance output presented in Figure 12.3 (also discussed in detail in Section 7.3), we observe that:

- There are some slight differences between the overall and balanced error rates (BER) with BER > overall, suggesting that minority classes might be ignored from the classification

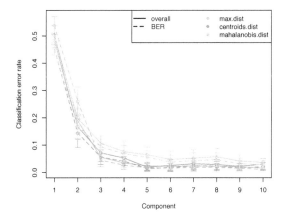

FIGURE 12.3: Tuning the number of components in PLS-DA on the SRBCT gene expression data. For each component, repeated cross-validation (10 × 3-fold CV) is used to evaluate the PLS-DA classification performance (overall and balanced error rate BER), for each type of prediction distance; `max.dist`, `centroids.dist` and `mahalanobis.dist`. Bars show the standard deviation across the repeated folds. The plot shows that the error rate reaches a minimum from three components.

task when considering the overall performance (`summary(Y)` shows that BL only includes eight samples). In general the trend is the same, however, and for further tuning with sPLS-DA we will consider the BER.

- The error rate decreases and reaches a minimum for `ncomp = 3` for the `max.dist` distance. These parameters will be included in further analyses.

Notes:

- *PLS-DA is an iterative model, where each component is orthogonal to the previous and gradually aims to build more discrimination between sample classes. We should always regard a final PLS-DA (with specified **ncomp**) as a 'compounding' model (i.e. PLS-DA with component 3 includes the trained model on the previous two components).*
- *We advise to use at least 50 repeats, and choose the number of folds that are appropriate for the sample size of the data set, as shown in Figure 12.3).*

Additional numerical outputs from the performance results are listed and can be reported as performance measures (not output here):

```
perf.plsda.srbct
```

12.5.3.3   Final PLS-DA model

We now run our final PLS-DA model that includes three components:

```
final.plsda.srbct <- plsda(X,Y, ncomp = 3)
```

We output the sample plots for the dimensions of interest (up to three). By default, the samples are coloured according to their class membership. We also add confidence ellipses (ellipse = TRUE, confidence level set to 95% by default, see the argument ellipse.level) in Figure 12.4. A 3D plot could also be insightful (use the argument type = '3D').

```
plotIndiv(final.plsda.srbct, ind.names = FALSE, legend=TRUE,
        comp=c(1,2), ellipse = TRUE,
        title = 'PLS-DA on SRBCT comp 1-2',
        X.label = 'PLS-DA comp 1', Y.label = 'PLS-DA comp 2')
```

```
plotIndiv(final.plsda.srbct, ind.names = FALSE, legend=TRUE,
        comp=c(2,3), ellipse = TRUE,
        title = 'PLS-DA on SRBCT comp 2-3',
        X.label = 'PLS-DA comp 2', Y.label = 'PLS-DA comp 3')
```

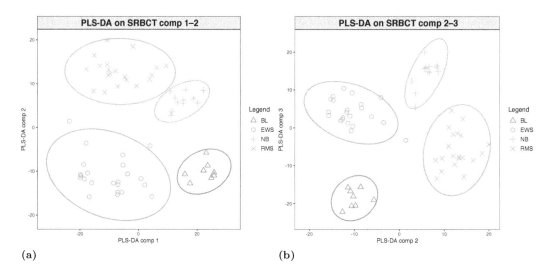

(a)   (b)

FIGURE 12.4: Sample plots from PLS-DA performed on the SRBCT gene expression data. Samples are projected into the space spanned by the first three components. (a) Components 1 and 2 and (b) components 1 and 3. Samples are coloured by their tumour subtypes. Component 1 discriminates RMS + EWS vs. NB + BL, component 2 discriminates RMS + NB vs. EWS + BL, while component 3 discriminates further the NB and BL groups. It is the combination of all three components that enables us to discriminate all classes.

We can observe improved clustering according to tumour subtypes, compared with PCA. This is to be expected since the PLS-DA model includes the class information of each sample. We observe some discrimination between the NB and BL samples vs. the others on the first component (x-axis), and EWS and RMS vs. the others on the second component (y-axis). From the plotIndiv() function, the axis labels indicate the amount of variation explained per component. However, the interpretation of this amount is *not as important* as in PCA, as PLS-DA aims to maximise the covariance between components associated to X and Y, rather than the variance X.

12.5.3.4   Classification performance

We can rerun a more extensive performance evaluation with more repeats for our final model:

```
set.seed(30) # For reproducibility with this handbook, remove otherwise
perf.final.plsda.srbct <- perf(final.plsda.srbct, validation = 'Mfold',
                               folds = 3,
                               progressBar = FALSE, # TRUE to track progress
                               nrepeat = 50)
```

Retaining only the BER and the `max.dist`, numerical outputs of interest include the final overall performance for three components:

```
perf.final.plsda.srbct$error.rate$BER[, 'max.dist']
```

```
##    comp1    comp2    comp3
## 0.54945 0.24640 0.06015
```

As well as the error rate per class across each component:

```
perf.final.plsda.srbct$error.rate.class$max.dist
```

```
##        comp1   comp2  comp3
## EWS 0.2348 0.08609 0.1026
## BL  0.7900 0.50250 0.0000
## NB  0.3750 0.33000 0.0500
## RMS 0.7980 0.06700 0.0880
```

From this output, we can see that the first component tends to classify EWS and NB better than the other classes. As components 2 and then 3 are added, the classification improves for all classes. However we see a slight increase in classification error in component 3 for EWS and RMS while BL is perfectly classified. A permutation test could also be conducted to conclude about the significance of the differences between sample groups, but is not currently implemented in the package.

12.5.3.5 Background prediction

A prediction background can be added to the sample plot by calculating a background surface first, before overlaying the sample plot (Figure 12.5, see Appendix 12.A, or `?background.predict`). We give an example of the code below based on the maximum prediction distance:

```
background.max <- background.predict(final.plsda.srbct,
                                     comp.predicted = 2,
                                     dist = 'max.dist')

plotIndiv(final.plsda.srbct, comp = 1:2, group = srbct$class,
          ind.names = FALSE, title = 'Maximum distance',
          legend = TRUE,  background = background.max)
```

Figure 12.5 shows the differences in prediction according to the prediction distance, and can be used as a further diagnostic tool for distance choice. It also highlights the characteristics of

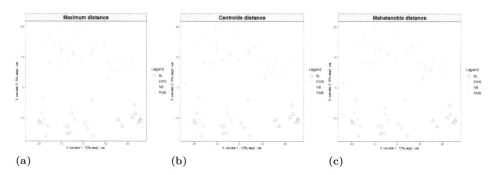

(a)   (b)   (c)

FIGURE 12.5: Sample plots from PLS-DA on the SRBCT gene expression data and prediction areas based on prediction distances. From our usual sample plot, we overlay a background prediction area based on permutations from the first two PLS-DA components using the three different types of prediction distances. The outputs show how the prediction distance can influence the quality of the prediction, with samples projected into a wrong class area, and hence resulting in predicted misclassification. *Note: currently, the prediction area background can only be calculated for the first two components.*

the distances. For example the `max.dist` is a linear distance, whereas both `centroids.dist` and `mahalanobis.dist` are non-linear. Our experience has shown that as discrimination of the classes becomes more challenging, the complexity of the distances (from maximum to Mahalanobis distance) should increase, see details in Appendix 12.A.

12.5.4   Example: sPLS-DA

In high-throughput experiments, we expect that many of the 2,308 genes in X are noisy or uninformative to characterise the different classes. An sPLS-DA analysis will help refine the sample clusters and select a small subset of variables relevant to discriminate each class.

12.5.4.1   Number of variables to select

We estimate the classification error rate with respect to the number of selected variables in the model with the function `tune.splsda()`. The tuning is being performed one component at a time inside the function and the optimal number of variables to select is automatically retrieved after each component run, as described in Section 7.3.

Previously, we determined the number of components to be `ncomp = 3` with PLS-DA. Here we set `ncomp = 4` to further assess if this would be the case for a sparse model, and use five-fold cross validation repeated ten times. We also choose the maximum prediction distance.

Note:

- *For a thorough tuning step, the following code should be repeated 10 to 50 times and the error rate is averaged across the runs. You may obtain slightly different results below for this reason.*

We first define a grid of `keepX` values. For example here, we define a fine grid at the start, and then specify a coarser, larger sequence of values:

```
# Grid of possible keepX values that will be tested for each comp
list.keepX <- c(1:10, seq(20, 100, 10))
list.keepX
```

```
## [1]   1   2   3   4   5   6   7   8   9  10  20  30
## [13]  40  50  60  70  80  90 100
```

```
# This chunk takes ~ 2 min to run
# Some convergence issues may arise but it is ok as this is run on CV folds
tune.splsda.srbct <- tune.splsda(X, Y, ncomp = 4, validation = 'Mfold',
                                 folds = 5, dist = 'max.dist',
                                 test.keepX = list.keepX, nrepeat = 10)
```

The following command line will output the mean error rate for each component and each tested `keepX` value given the past (tuned) components.

```
# Just a head of the classification error rate per keepX (in rows) and comp
head(tune.splsda.srbct$error.rate)
```

```
##     comp1  comp2   comp3   comp4
## 1 0.6263 0.2968 0.07664 0.01548
## 2 0.5700 0.2959 0.05585 0.01782
## 3 0.5569 0.2880 0.04273 0.01782
## 4 0.5347 0.2848 0.03414 0.01673
## 5 0.5245 0.2787 0.02893 0.01673
## 6 0.5253 0.2763 0.02560 0.01564
```

When we examine each individual row, this output globally shows that the classification error rate continues to decrease after the third component in sparse PLS-DA.

We display the mean classification error rate on each component, bearing in mind that each component is conditional on the previous components calculated with the optimal number of selected variables. The diamond in Figure 12.6 indicates the best `keepX` value to achieve the lowest error rate per component.

```
# To show the error bars across the repeats:
plot(tune.splsda.srbct, sd = TRUE)
```

The tuning results depend on the tuning grid `list.keepX`, as well as the values chosen for `folds` and `nrepeat`. Therefore, we recommend assessing the performance of the *final* model, as well as examining the stability of the selected variables across the different folds, as detailed in the next section.

Figure 12.6 shows that the error rate decreases when more components are included in sPLS-DA. To obtain a more reliable estimation of the error rate, the number of repeats should be increased (between 50 and 100). This type of graph helps not only to choose the 'optimal' number of variables to select, but also to confirm the number of components `ncomp`. From the code below, we can assess that in fact, the addition of a fourth component does not improve the classification (no statistically significant improvement according to a one-sided t-test), hence we can choose `ncomp = 3`.

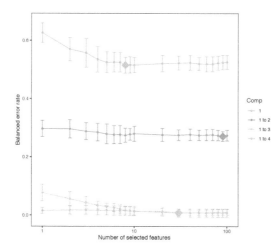

FIGURE 12.6: Tuning `keepX` for the sPLS-DA performed on the SRBCT gene expression data. Each coloured line represents the balanced error rate (y-axis) per component across all tested `keepX` values (x-axis) with the standard deviation based on the repeated cross-validation folds. The diamond indicates the optimal `keepX` value on a particular component which achieves the lowest classification error rate as determined with a one-sided *t*−test. As sPLS-DA is an iterative algorithm, values represented for a given component (e.g. comp 1 to 2) include the optimal `keepX` value chosen for the previous component (comp 1).

```
# The optimal number of components according to our one-sided t-tests
tune.splsda.srbct$choice.ncomp$ncomp
```

```
## [1] 3
```

```
# The optimal keepX parameter according to minimal error rate
tune.splsda.srbct$choice.keepX
```

```
## comp1 comp2 comp3 comp4
##     8    90    30    30
```

12.5.4.2   Final model and performance

Here is our final sPLS-DA model with three components and the optimal `keepX` obtained from our tuning step.

You can choose to skip the tuning step, and input your arbitrarily chosen parameters in the following code (simply specify your own **ncomp** and **keepX** values):

```
# Optimal number of components based on t-tests on the error rate
ncomp <- tune.splsda.srbct$choice.ncomp$ncomp
ncomp
```

```
## [1] 3
```

```
# Optimal number of variables to select
select.keepX <- tune.splsda.srbct$choice.keepX[1:ncomp]
select.keepX
```

```
## comp1 comp2 comp3
##     8    90    30
```

```
splsda.srbct <- splsda(X, Y, ncomp = 3, keepX = select.keepX)
```

The performance of the model with the **ncomp** and **keepX** parameters is assessed with the **perf()** function. We use five-fold validation (**folds = 5**), repeated ten times (**nrepeat = 10**) for illustrative purposes, but we recommend increasing to **nrepeat = 50**. Here we choose the **max.dist** prediction distance, based on our results obtained with PLS-DA.

The classification error rates that are output include both the overall error rate, as well as the balanced error rate (BER) when the number of samples per group is not balanced – as is the case in this study.

```
set.seed(34)   # For reproducibility with this handbook, remove otherwise
```

```
perf.splsda.srbct <- perf(splsda.srbct, folds = 5, validation = "Mfold",
                dist = "max.dist", progressBar = FALSE, nrepeat = 10)
```

```
# perf.splsda.srbct  # Lists the different outputs
perf.splsda.srbct$error.rate
```

```
## $overall
##          max.dist
## comp1   0.43651
## comp2   0.21429
## comp3   0.01111
##
## $BER
##          max.dist
## comp1   0.52069
## comp2   0.28331
## comp3   0.01405
```

We can also examine the error rate per class:

```
perf.splsda.srbct$error.rate.class
```

```
## $max.dist
##         comp1    comp2     comp3
## EWS  0.02609  0.01739  0.008696
## BL   0.57500  0.36250  0.037500
## NB   0.91667  0.60833  0.000000
## RMS  0.56500  0.14500  0.010000
```

These results can be compared with the performance of PLS-DA and show the benefits of variable selection to not only obtain a parsimonious model but also to improve the classification error rate (overall and per class).

12.5.4.3 Variable selection and stability

During the repeated cross-validation process in `perf()` we can record how often the same variables are selected across the folds. This information is important to answer the question: *How reproducible is my gene signature when the training set is perturbed via cross-validation?*.

```
par(mfrow=c(1,2))
# For component 1
stable.comp1 <- perf.splsda.srbct$features$stable$comp1
barplot(stable.comp1, xlab = 'variables selected across CV folds',
        ylab = 'Stability frequency',
        main = 'Feature stability for comp = 1')

# For component 2
stable.comp2 <- perf.splsda.srbct$features$stable$comp2
barplot(stable.comp2, xlab = 'variables selected across CV folds',
        ylab = 'Stability frequency',
        main = 'Feature stability for comp = 2')
par(mfrow=c(1,1))
```

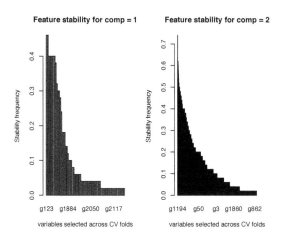

FIGURE 12.7: Stability of variable selection from the sPLS-DA on the SRBCT gene expression data. We use a by-product from `perf()` to assess how often the same variables are selected for a given `keepX` value in the final sPLS-DA model. The barplot represents the frequency of selection across repeated CV folds for each selected gene for components 1 and 2. The genes are ranked according to decreasing frequency.

Figure 12.7 shows that the genes selected on component 1 are moderately stable (frequency < 0.5) whereas those selected on component 2 are more stable (frequency < 0.7). This can be explained as there are various combinations of genes that are discriminative on component 1, whereas the number of combinations decreases as we move to component 2 which attempts to refine the classification.

The function `selectVar()` outputs the variables selected for a given component and their loading values (ranked in decreasing absolute value). We concatenate those results with the feature stability, as shown here for variables selected on component 1:

```
# First extract the name of selected var:
select.name <- selectVar(splsda.srbct, comp = 1)$name

# Then extract the stability values from perf:
stability <- perf.splsda.srbct$features$stable$comp1[select.name]

# Just the head of the stability of the selected var:
head(cbind(selectVar(splsda.srbct, comp = 1)$value, stability))
```

```
##          value.var  Var1 Freq
## g123        0.6639  g123 0.46
## g846        0.4519  g846 0.46
## g1606       0.3015 g1606 0.30
## g335        0.2954  g335 0.30
## g836        0.2569  g836 0.40
## g783        0.2110  g783 0.24
```

As we hinted previously, the genes selected on the first component are not necessarily the most stable, suggesting that different combinations can lead to the same discriminative ability of the model. The stability increases in the following components, as the classification task becomes more refined.

Note:

- *You can also apply the vip() function on splsda.srbct.*

12.5.4.4 Sample visualisation

Previously, we showed the ellipse plots displayed for each class. Here we also use the star argument (`star = TRUE`), which displays arrows starting from each group centroid towards each individual sample (Figure 12.8).

```
plotIndiv(splsda.srbct, comp = c(1,2),
          ind.names = FALSE,
          ellipse = TRUE, legend = TRUE,
          star = TRUE,
          title = 'SRBCT, sPLS-DA comp 1 - 2')


plotIndiv(splsda.srbct, comp = c(2,3),
          ind.names = FALSE,
          ellipse = TRUE, legend = TRUE,
          star = TRUE,
          title = 'SRBCT, sPLS-DA comp 2 - 3')
```

The sample plots are different from PLS-DA (Figure 12.4) with an overlap of specific classes (i.e. NB + RMS on components 1 and 2), that are then further separated on component

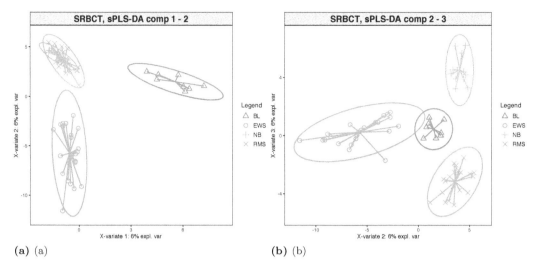

(a) (a)   **(b)** (b)

FIGURE 12.8: Sample plots from the sPLS-DA performed on the SRBCT gene expression data. Samples are projected into the space spanned by the first three components. The plots represent 95% ellipse confidence intervals around each sample class. The start of each arrow represents the centroid of each class in the space spanned by the components. (a) Components 1 and 2 and (b) components 2 and 3. Samples are coloured by their tumour subtype. Component 1 discriminates BL vs. the rest, component 2 discriminates EWS vs. the rest, while component 3 further discriminates NB vs. RMS vs. the rest. The combination of all three components enables us to discriminate all classes.

3, thus showing how the genes selected on each component discriminate particular sets of sample groups.

12.5.4.5   Variable visualisation

We represent the genes selected with sPLS-DA on the correlation circle plot. Here to increase interpretation, we specify the argument `var.names` as the first 10 characters of the gene names (Figure 12.9). We also reduce the size of the font with the argument `cex`.

Note:

- *We can store the plotvar() as an object to output the coordinates and variable names if the plot is too cluttered.*

```
var.name.short <- substr(srbct$gene.name[, 2], 1, 10)
plotVar(splsda.srbct, comp = c(1,2),
        var.names = list(var.name.short), cex = 3)
```

By considering both the correlation circle plot (Figure 12.9) and the sample plot in Figure 12.8, we observe that a group of genes with a positive correlation with component 1 ('EH domain', 'proteasome' etc.) are associated with the BL samples. We also observe two groups of genes either positively or negatively correlated with component 2. These genes are likely to characterise either the NB + RMS classes, or the EWS class. This interpretation can be further examined with the `plotLoadings()` function.

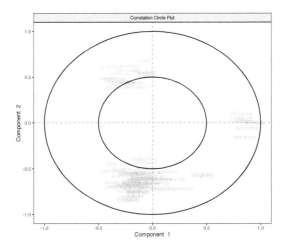

FIGURE 12.9: Correlation circle plot representing the genes selected by sPLS-DA performed on the SRBCT gene expression data. Gene names are truncated to the first 10 characters. Only the genes selected by sPLS-DA are shown in components 1 and 2. We observe three groups of genes (positively associated with component 1, and positively or negatively associated with component 2). This graphic should be interpreted in conjunction with the sample plot.

In the plot shown in Figure 12.10, the loading weights of each selected variable on each component are represented (see Section 6.2). The colours indicate the group in which the expression of the selected gene is maximal based on the mean (`method = 'median'` is also available for skewed data). For example on component 1:

```
plotLoadings(splsda.srbct, comp = 1, method = 'mean', contrib = 'max',
             name.var = var.name.short)
```

Here all genes are associated with BL (on average, their expression levels are higher in this class than in the other classes).

Notes:

- *Consider using the argument **ndisplay** to only display the top selected genes if the signature is too large.*
- *Consider using the argument **contrib = 'min'** to interpret the inverse trend of the signature (i.e. which genes have the smallest expression in which class, here a mix of NB and RMS samples).*

To complete the visualisation, the CIM in this special case is a simple hierarchical heatmap (see `?cim`) representing the expression levels of the genes selected across all three components with respect to each sample. Here we use an Euclidean distance with Complete agglomeration method, and we specify the argument `row.sideColors` to colour the samples according to their tumour type (Figure 12.11).

```
cim(splsda.srbct, row.sideColors = color.mixo(Y))
```

The heatmap shows the level of expression of the genes selected by sPLS-DA across all

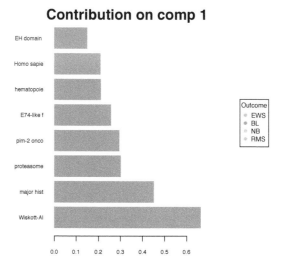

FIGURE 12.10: Loading plot of the genes selected by sPLS-DA on component 1 on the SRBCT gene expression data. Genes are ranked according to their loading weight (most important at the bottom to least important at the top), represented as a barplot. Colours indicate the class for which a particular gene is maximally expressed, on average, in this particular class. The plot helps to further characterise the gene signature and should be interpreted jointly with the sample plot (Figure 12.8).

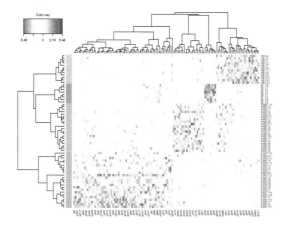

FIGURE 12.11: Clustered Image Map of the genes selected by sPLS-DA on the SRBCT gene expression data across all three components. A hierarchical clustering based on the gene expression levels of the selected genes, with samples in rows coloured according to their tumour subtype (using Euclidean distance with Complete agglomeration method). As expected, we observe a separation of all different tumour types, which are characterised by different levels of expression.

three components, and the overall ability of the gene signature to discriminate the tumour subtypes.

Note:

- *You can change the argument* `comp` *if you wish to visualise a specific set of components in* `cim()`.

12.5.5   Take a detour: Prediction

In this section, we artificially create an 'external' test set on which we want to predict the class membership to illustrate the prediction process in sPLS-DA (see Appendix 12.A). We randomly select 50 samples from the `srbct` study as part of the training set, and the remainder as part of the test set:

```
set.seed(33) # For reproducibility with this handbook, remove otherwise
train <- sample(1:nrow(X), 50)    # Randomly select 50 samples in training
test <- setdiff(1:nrow(X), train) # Rest is part of the test set

# Store matrices into training and test set:
X.train <- X[train, ]
X.test <- X[test,]
Y.train <- Y[train]
Y.test <- Y[test]

# Check dimensions are OK:
dim(X.train); dim(X.test)
```

```
## [1]    50 2308
## [1]    13 2308
```

Here we assume that the tuning step was performed on the training set *only* (it is *really important* to tune only on the training step to avoid overfitting), and that the optimal `keepX` values are, for example, `keepX = c(20,30,40)` on three components. The final model on the training data is:

```
train.splsda.srbct <- splsda(X.train, Y.train, ncomp = 3, keepX = c(20,30,40))
```

We now apply the trained model on the test set `X.test` and we specify the prediction distance, for example `mahalanobis.dist` (see also `?predict.splsda`):

```
predict.splsda.srbct <- predict(train.splsda.srbct, X.test,
                        dist = "mahalanobis.dist")
```

The `$class` output of our object `predict.splsda.srbct` gives the predicted classes of the test samples.

First we concatenate the prediction for each of the three components (conditionally on the previous component) and the real class - in a real application case you may not know the true class.

```
# Just the head:
head(data.frame(predict.splsda.srbct$class, Truth = Y.test))
```

```
##          mahalanobis.dist.comp1 mahalanobis.dist.comp2
## EWS.T7                     EWS                    EWS
## EWS.T15                    EWS                    EWS
## EWS.C8                     EWS                    EWS
## EWS.C10                    EWS                    EWS
## BL.C8                       BL                     BL
## NB.C6                       NB                     NB
##          mahalanobis.dist.comp3 Truth
## EWS.T7                     EWS   EWS
## EWS.T15                    EWS   EWS
## EWS.C8                     EWS   EWS
## EWS.C10                    EWS   EWS
## BL.C8                       BL    BL
## NB.C6                       NB    NB
```

If we only look at the final prediction on component 2, compared to the real class:

```
# Compare prediction on the second component and change as factor
predict.comp2 <- predict.splsda.srbct$class$mahalanobis.dist[,2]
table(factor(predict.comp2, levels = levels(Y)), Y.test)
```

```
##       Y.test
##       EWS BL NB RMS
##   EWS   4  0  0   0
##   BL    0  1  0   0
##   NB    0  0  1   1
##   RMS   0  0  0   6
```

And on the third component:

```
# Compare prediction on the third component and change as factor
predict.comp3 <- predict.splsda.srbct$class$mahalanobis.dist[,3]
table(factor(predict.comp3, levels = levels(Y)), Y.test)
```

```
##       Y.test
##       EWS BL NB RMS
##   EWS   4  0  0   0
##   BL    0  1  0   0
##   NB    0  0  1   0
##   RMS   0  0  0   7
```

The prediction is better on the third component, compared to a two-component model.

Next, we look at the output $predict, which gives the predicted dummy scores assigned for each test sample and each class level for a given component (as explained in Appendix 12.A). Each column represents a class category:

```
# On component 3, just the head:
head(predict.splsda.srbct$predict[, , 3])
```

```
##              EWS       BL       NB      RMS
## EWS.T7   1.26849 -0.05274 -0.24071  0.024961
## EWS.T15  1.15058 -0.02222 -0.11878 -0.009583
## EWS.C8   1.25628  0.05481 -0.16500 -0.146093
## EWS.C10  0.83996  0.10871  0.16453 -0.113200
## BL.C8    0.02431  0.90877  0.01775  0.049163
## NB.C6    0.06738  0.05087  0.86247  0.019275
```

In PLS-DA and sPLS-DA, the final prediction call is given based on this matrix on which a pre-specified distance (such as `mahalanobis.dist` here) is applied. From this output, we can understand the link between the dummy matrix Y, the prediction, and the importance of choosing the prediction distance. More details are provided in Appendix 12.A.

12.5.6   AUROC outputs complement performance evaluation

As PLS-DA acts as a classifier, we can plot the AUC (Area Under The Curve) ROC (Receiver Operating Characteristics) to complement the sPLS-DA classification performance results (see Section 7.3). The AUC is calculated from training cross-validation sets and averaged. The ROC curve is displayed in Figure 12.12. In a multiclass setting, each curve represents one class vs. the others and the AUC is indicated in the legend, and also in the numerical output:

```
auc.srbct <- auroc(splsda.srbct)
```

```
## $Comp1
##                   AUC   p-value
## EWS vs Other(s) 0.3902 1.493e-01
## BL vs Other(s)  1.0000 5.586e-06
## NB vs Other(s)  0.8105 8.821e-04
## RMS vs Other(s) 0.6523 5.308e-02
##
## $Comp2
##                   AUC   p-value
## EWS vs Other(s) 1.0000 5.135e-11
## BL vs Other(s)  1.0000 5.586e-06
## NB vs Other(s)  0.8627 1.020e-04
## RMS vs Other(s) 0.8140 6.699e-05
##
## $Comp3
##                 AUC   p-value
## EWS vs Other(s)   1 5.135e-11
## BL vs Other(s)    1 5.586e-06
## NB vs Other(s)    1 8.505e-08
## RMS vs Other(s)   1 2.164e-10
```

The ideal ROC curve should be along the top left corner, indicating a high true positive

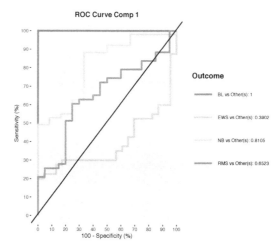

FIGURE 12.12: ROC curve and AUC from sPLS-DA on the SRBCT gene expression data on component 1 averaged across one-vs.-all comparisons. Numerical outputs include the AUC and a Wilcoxon test p-value for each 'one vs. other' class comparisons that are performed per component. This output complements the sPLS-DA performance evaluation but *should not be used for tuning* (as the prediction process in sPLS-DA is based on prediction distances, not a cutoff that maximises specificity and sensitivity as in ROC). The plot suggests that the sPLS-DA model can distinguish BL subjects from the other groups with a high true positive and low false positive rate, while the model is less well able to distinguish samples from other classes on component 1.

rate (sensitivity on the y-axis) and a high true negative rate (or low 100 – specificity on the x-axis), with an AUC close to 1. This is the case for BL vs. the others on component 1. The numerical output shows a perfect classification on component 3.

Note:

- *A word of caution when using the ROC and AUC in s/PLS-DA: these criteria may not be particularly insightful, or may not be in full agreement with the s/PLS-DA performance, as the prediction threshold in PLS-DA is based on a specified distance as we described earlier in this Section and in Appendix 12.A. Thus, such a result complements the sPLS-DA performance we have calculated earlier in Section 12.5.*

12.6   To go further

12.6.1   Microbiome

Like any other approach in `mixOmics`, PLS-DA and sPLS-DA can be applied to microbiome data. The data from 16S rRNA amplicon sequencing and shotgun metagenomics are characterised by their sparse and compositional nature (Section 4.2). We recommend using a centered log ratio transformation (`logratio = CLR`) in `plsda()` or `splsda()` to project the compositional data into a Euclidean space. More details are available in Lê Cao et al. (2016) and detailed code on our website `http://mixomics.org/mixmc/`.

12.6.2 Multilevel

In the case of a repeated measurement experimental design (see Section 4.1), often the individual variation (on which all treatments or conditions are applied) can mask the more subtle treatment effect. The multilevel decomposition can apply to a supervised framework (Liquet et al., 2012). We give here two brief examples.

12.6.2.1 Example: `vac18`

In the `vac18` gene expression study, the cells from the same patient undergo different treatments (see `?vac18`). However, the interest is in discriminating the vaccine stimulation type:

```
data(vac18)
X <- vac18$genes
# The treatments
summary(vac18$stimulation)
```

```
## LIPO5  GAG+  GAG-    NS
##    11    10    10    11
```

```
# The number of repeated measurements per unique individual
summary(as.factor(vac18$sample))
```

```
##  1  2  3  4  5  6  7  8  9 10 11 12
##  4  4  2  4  3  3  4  3  4  4  4  3
```

For a PLS-DA model we specify the information about the repeated measurements (e.g. the unique ID of each patient) as:

```
plsda.vac18 <- plsda(X, Y = vac18$stimulation, multilevel = vac18$sample,
                     ncomp = 3)
```

A detailed vignette can be found in http://mixomics.org/case-studies/multilevel-vac18/.

12.6.2.2 Example: `diverse.16S`

In this 16S rRNA gene amplicon data set, the microbiome is sequenced in different habitats (three bodysites) in the same individuals. We are interested in discriminating the bodysites. In addition, we also transform the count data using centered log ratio transformation:

```
data('diverse.16S')
# The data are filtered raw counts
X <- diverse.16S$data.raw
```

```
# The outcome
Y <- diverse.16S$bodysite
```

```
# Unique ID of each individual for multilevel analysis and summary
#summary(as.factor(diverse.16S$sample))
dim(X)
```

```
## [1]   162 1674
```

The PLS-DA model is:

```
plsda.16S <- plsda(X, Y = diverse.16S$bodysite,
                   multilevel = diverse.16S$sample, logratio = 'CLR',
                   ncomp = 3)
```

A detailed vignette can be found in http://mixomics.org/mixmc/case-study-hmp-bodysites-repeated-measures/ if you wish to pursue the analysis (otherwise, use the quick-start functions described in 12.5).

12.6.3   Other related methods and packages

Other PLS-based discriminant analyses exist, such as Orthogonal PLS-DA (OPLS-DA, Bylesjö et al. (2006), see also Section 10.7), which was reported to have the same discriminant power as PLS-DA (Tapp and Kemsley, 2009) and is also suited for class prediction of unknown samples. The packages `ropls` and `muma` (dedicated to the analysis of metabolomics data) implement PLS-DA (Thévenot et al., 2015) and OPLS-DA (Gaude et al., 2013).

12.7   FAQ

Can I discriminate more than two groups of samples (multi-class classification)?

Yes, this is one of the advantages of PLS-DA.

Can I have a hierarchy between two factors (e.g. diet nested into genotype)?

Unfortunately no, sparse PLS-DA only allows us to discriminate all groups at once (i.e. 4×2 groups when there are four diets and two genotypes).

Can I have missing values in my data?

In the X data set, yes, including when using the prediction functions `tune()`, `perf()`, and `predict()`. In the y factor, no, but samples that are missing their class information can be used as an external data for class prediction (without having the ground truth for these samples).

Can I skip the tuning step?

Sometimes having a small number of samples does not allow for cross-validation or leave-one-out cross-validation. In that case, an exploratory approach can be adopted, where `ncomp`, and `keepX` are chosen *apriori*, and the steps 1–2 from the framework presented in Section

12.3 can be omitted. The final models can still be obtained, as well as the performance evaluation with `perf()`, however, caution must be taken when interpreting the results.

12.8 Summary

PLS-DA is a supervised framework useful for analysing a data set where we wish to select variables that will correctly predict samples of known classification, as well as predict the class of new samples. PLS-DA can be seen as a special case of Linear Discriminant Analysis (LDA). It constructs linear combinations of variables to discriminate sample groups informed by a categorical outcome variable (e.g. cancer subtype), which is coded into a dummy matrix. Contrary to LDA, PLS-DA only takes into account between group variability rather than within and between group variability, but is suitable for multicollinear problems from large data sets.

PLS-DA calculates artificial components from the original variables to reduce the data dimensions whilst maximising the separation between sample classes. sparse PLS-DA can then be employed to select a subset of variables which best discriminate the outcome.

PLS-DA is prone to overfitting, whereby the variables selected discriminate the classes of samples very well, but do not generalise well to new data sets. With `mixOmics` we propose to use cross-validation to not only tune the 'best' number of components and variables to select on each component, but also to assess the performance of the method, along with diagnostic graphical outputs.

12.A Appendix: Prediction in PLS-DA

12.A.1 Prediction distances

Different prediction distances are proposed and implemented in the functions `predict()`, `tune()` and `perf()` to assign to each new observation a final predicted class.

Mathematically, we can define those predicted outputs for a model with H components as follows. Recall that the outcome matrix $\boldsymbol{Y}$ is a dummy matrix of size $(N \times K)$. For a new data matrix $\boldsymbol{X}_{\text{new}}$ of size $(N_{\text{new}} \times P)$, we define the *predicted dummy variables* $\widehat{\boldsymbol{Y}}_{\text{new}}$ of size $(N_{\text{new}} \times K)$ as:

$$\widehat{\boldsymbol{Y}}_{\text{new}} = \boldsymbol{X}_{\text{new}} \boldsymbol{W} (\boldsymbol{D}^T \boldsymbol{W})^{-1} \boldsymbol{B},$$

where $\boldsymbol{W}, \boldsymbol{D}$ and $\boldsymbol{B}$ are derived from the training data sets $\boldsymbol{X}$ and $\boldsymbol{Y}$. $\boldsymbol{W}$ is a $(P \times H)$ matrix containing the loading vectors associated to $\boldsymbol{X}$, $\boldsymbol{D}$ is a $(P \times H)$ matrix containing the regression coefficients of $\boldsymbol{X}$ on its H latent components and $\boldsymbol{B}$ is a $(H \times K)$ matrix containing the regression coefficients of $\boldsymbol{Y}$ on the H latent components associated to $\boldsymbol{X}$. Therefore, $\widehat{\boldsymbol{Y}}_{\text{new}}$ is the prediction from a multivariable (several columns in $\boldsymbol{Y}_{\text{new}}$) multivariate (several predictors in $\boldsymbol{X}$) model.

We define the *predicted scores* or *predicted components* $\boldsymbol{T}_{\text{pred}}$ of size $(N_{\text{new}} \times H)$ as:

$$\boldsymbol{T}_{\text{pred}} = \boldsymbol{X}_{\text{new}}\boldsymbol{W}(\boldsymbol{D}^T\boldsymbol{W})^{-1}$$

with the same notations as above. The prediction distances are then applied as follows:

The maximum distance `max.dist` is applied to the predicted dummy values $\widehat{\boldsymbol{Y}}_{\text{new}}$. This distance represents the most intuitive method to predict the class of a new observation sample, as the predicted class is the outcome category with the largest predicted dummy value. The distance performs well in single data set analysis with multiclass problems (Lê Cao et al., 2011).

For the centroid-based distances Mahalanobis and Centroids, we first calculate the centroid $\boldsymbol{G}_k$ of all training samples belonging to the class $k \leq K$ based on the H latent components associated to $\boldsymbol{X}$. Both Mahalanobis and Centroids distances are applied on the predicted scores $\boldsymbol{T}_{\text{pred}}$. The predicted class of a new observation is

$$\operatorname*{argmin}_{1 \leq k \leq K} \left\{ \operatorname{dist}(\boldsymbol{T}_{\text{pred}}, \boldsymbol{G}_k) \right\}, \tag{12.3}$$

i.e. the class for which the distance between its centroid and the H predicted scores is minimal. Denote $\boldsymbol{t}_{\text{pred}}^h$, a predicted component for a given dimension h, we define the following distances:

- The Centroids distance, which solves Equation (12.3) using the Euclidean distance $\operatorname{dist}(\boldsymbol{t}_{\text{pred}}^h, \boldsymbol{G}_k) = \sqrt{\sum_{h=1}^{H} \left(\boldsymbol{t}_{\text{pred}}^h - \boldsymbol{G}_k^h \right)^2}$, where $\boldsymbol{G}_k^h$ denotes the centroid of all training samples from class k based on latent component h.

- The Mahalanobis distance, which solves Equation (12.3) using the Mahalanobis distance $\operatorname{dist}(\boldsymbol{t}_{\text{pred}}^h, \boldsymbol{G}_k) = \sqrt{(\boldsymbol{t}_{\text{pred}}^h - \boldsymbol{G}_k)^T \boldsymbol{S}^{-1}(\boldsymbol{t}_{\text{pred}}^h - \boldsymbol{G}_k)}$, where $\boldsymbol{S}$ is the variance-covariance matrix of $\boldsymbol{t}_{\text{pred}}^h - \boldsymbol{G}_k$.

In practice we found that the centroid-based distances, and specifically the Mahalanobis distance led to more accurate predictions than the maximum distance for complex classification problems and N−integration problems. The centroid distances consider the prediction in a H dimensional space using the predicted scores, while the maximum distance considers a single point estimate using the predicted dummy variables on the last dimension of the model. We can assess the different distances and choose the prediction distance that achieves the best performance using the `tune()` and `perf()` outputs.

In Table 12.2, we take the same example from SRBCT as in Section 12.5 and output both the predicted components $\boldsymbol{T}_{\text{pred}}$ and the prediction call from `max.dist` on the new samples:

```
predict.splsda.srbct2 <- predict(train.splsda.srbct, X.test, dist = "max.dist")
```

```
data.frame(predict.splsda.srbct2$variates,
           max.dist = predict.splsda.srbct2$class$max.dist)
```

The centroids for each class can also be extracted (from the training data set), see Table 12.3:

TABLE 12.2: Example of predicted components for each dummy variable corresponding to each class, and predicted class based on maximum distance on test samples from the SRBCT data. The true class of the test samples is indicated in the row names.

	dim1	dim2	dim3	max.dist.comp1	max.dist.comp2	max.dist.comp3
EWS.T7	5.3564	0.7366	−1.2988	EWS	EWS	EWS
EWS.T15	4.5859	0.0240	−0.8289	EWS	EWS	EWS
EWS.C8	5.0895	−0.8772	−1.6927	EWS	EWS	EWS
EWS.C10	2.5255	−1.9571	−0.1822	EWS	EWS	EWS
BL.C8	−2.6765	−2.9535	−6.8453	RMS	NB	BL
NB.C6	−2.1094	−3.3371	4.2918	RMS	NB	NB
RMS.C4	−2.0997	3.9255	0.9367	RMS	RMS	RMS
RMS.C2	−0.6469	1.8502	1.6714	RMS	RMS	RMS
RMS.C8	−2.0171	1.3223	1.2797	RMS	RMS	RMS
RMS.C10	−2.3192	1.4922	1.6336	RMS	RMS	RMS
RMS.T6	−0.7311	3.7377	1.6683	RMS	RMS	RMS
RMS.T8	−0.9920	4.9827	−0.0596	RMS	RMS	RMS
RMS.T10	1.0621	4.8378	0.6251	EWS	RMS	RMS

TABLE 12.3: Example of centroids coordinates for each class from the training set, based on the example from the SRBCT data.

	dim1	dim2	dim3
EWS	3.337	−0.5544	−0.1942
BL	−2.649	−3.2042	−7.5601
NB	−2.058	−2.8437	4.8771
RMS	−1.710	4.9418	0.2279

```
predict.splsda.srbct2$centroids
```

12.A.2 Background area

The `background.predict()` function simulates points within a rectangle defined by the arguments `xlim` on the x-axis and `ylim` on the y-axis. The number of points is defined by an area of `resolution*resolution` (set to 100×100 by default). The `xlim` and `ylim` arguments are set by default to 1.2 * range of the X associated components (t_1, t_2), respectively. On each point we predict their class based on their coordinates $(t_{1_{pred}}, t_{2_{pred}})$. The algorithm estimates the predicted area for each class, defined as a 2-D surface where all points are predicted to be of the same class. A polygon is returned and can then be passed to `plotIndiv()` for plotting the predicted background area.

13

$N-$data integration

13.1 Why use $N-$integration methods?

In this chapter, we broadly introduce our $N-$integration methods available in `mixOmics`, where experiments are made on the same samples, or observations (introduced in Section 4.3). The methods are all based on regularised or sparse generalised canonical correlation analysis (RGCCA and sGCCA from Tenenhaus and Tenenhaus (2011) and Tenenhaus et al. (2014)) to extend the methods PLS and PLS-DA presented in Chapters 10 and 12. As such, this chapter is the culmination of our integrative approaches.

To showcase these methods, we will focus on the multiblock sPLS-DA extension called DIABLO (Singh et al., 2019) to integrate a series of continuous data matrices, or *blocks*, $\boldsymbol{X}_1,\ldots,\boldsymbol{X}_Q$ with a categorical outcome variable $\boldsymbol{y}$ transformed into a dummy matrix (described in Section 12.2). In Section 13.6, we review other available frameworks for multi-block regression and unsupervised analyses.

The analytical aim of multiblock sPLS-DA is to identify a signature composed of highly correlated features across the different types of omics which also discriminates a given outcome (*a multi-omics signature*). Solving this problem is not straightforward and additional parameters, such as data set weights, are required.

The biological aims are:

- To understand the role of each omics in a biological system,
- To improve our understanding of the relationship between the omics types,
- To identify a correlated/co-expressed molecular signature leading to more insight into molecular mechanisms,
- To develop a predictive analytical model towards personalised medicine.

Multiblock sPLS-DA is a holistic approach with the potential to find new biological insights not revealed by any single-data omics analysis, as some pathways are common to all data types, while other pathways may be specific to particular data types, as we have shown in an infant multi-omics study in Lee et al. (2019).

13.1.1 Biological question

Can I discriminate samples across several data sets based on their outcome category? Which variables across the different omics data sets discriminate the different outcomes? Can they constitute a multi-omics signature that predicts the class of external samples?

DOI: 10.1201/9781003026860-13

13.1.2   Statistical point of view and analytical challenges

These questions consider more than two numerical data sets and a categorical outcome. We have referred to this framework as $N-$integration as every data set is measured on the same N samples (introduced in Section 4.3). To answer these questions, a generalised version of PLS is required. The selection of the most relevant variables in each data set can be achieved using a lasso penalised version of the method.

The aim of $N-$integration with our sparse methods is to identify correlated (or co-expressed) variables measured on heterogeneous data sets which also explain the categorical outcome of interest in a supervised analysis. This multiple data set integration task is not trivial, as the analysis can be strongly affected by the variation between manufacturers or omics technological platforms despite being measured on the same biological samples.

Before embarking on multi-omic data integration, we strongly suggest performing individual and paired analyses first, for example with sPLS and sPLS-DA (Chapters 10 and 12), to understand the major sources of variation in each data set. *Such preliminary analysis will guide the full integration process and should not be skipped.*

The complexity of the integrative method increases along with the complexity of the biological question (and the data). We will need to specify information about the assumed relationship between pairs of data sets in the statistical model. Other challenges for multi-block data integration include the large number of highly collinear variables, and, often, a poorly-defined biological question (*'I want to integrate my data'*). As discussed in the cycle of analysis in Section 2.1, defining a suitable question (often framed in terms of known biology) is a necessary step for choosing the best design. Finally, maximising correlation *and* discrimination in a single optimisation problem are two opposing tasks and compromise is needed, as we discuss in Singh et al. (2019) and later in this chapter.

13.2   Principle

13.2.1   Multiblock sPLS-DA

Multiblock sPLS-DA (DIABLO) extends sparse Generalised Canonical Correlation Analysis (Tenenhaus and Tenenhaus, 2011), which, contrary to what its name suggests, generalises PLS (Chapter 10) and PLS-DA (Chapter 12) for multiple, matching data sets. Starting from the RGCCA package, we further extended these methods to fit into mixomics. Several types of analyses are available, such as regression for $N-$integration (block.pls, block.spls) as further described in Section 13.6, and classification analyses (block.plsda, block.splsda), illustrated in this chapter.

We seek for latent components – linear combinations of variables, from each omics data set so that the sum of covariances between each pair of data sets, including the outcome, is maximised (Figure 13.1). The covariance of each pair of data sets is weighted according to a specified design matrix. Let Q denote omics data sets $X_1(N \times P_1), \dots, X_Q(N \times P_Q)$ (each data set includes a different type of variable, which vary in number). For the first dimension, the general sGCCA solves:

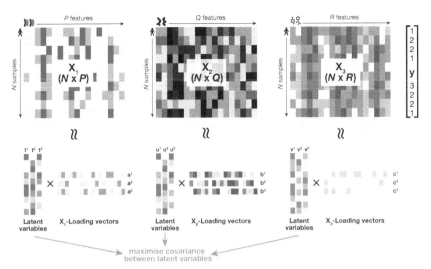

FIGURE 13.1: Schematic view of multiblock sPLS-DA (DIABLO) matrix decomposition of $X_1, \ldots, X_Q$ and Y into sets of latent variables (components) and loading vectors. The components are defined so that their paired covariance (defined by a *design* matrix that weights each data set pair, as detailed in Section 13.5) is maximised. The outcome y is internally transformed as a dummy matrix Y in the method.

$$\underset{a_1, \ldots, a_Q}{\operatorname{argmax}} \sum_{q,k=1, q \neq k}^{Q} c_{q,k} \ \operatorname{cov}(X_q a_q, X_k a_k),$$

$$\text{subject to} \quad ||a_q||_2 = 1 \text{ and } ||a_q||_1 \leq \lambda_q \text{ for all } 1 \leq q \leq Q$$

Let us have a closer look at Equation (13.1):

- $X_q a_q$ and $X_k a_k$ are linear combinations of the data sets X_q and X_k, respectively, weighted by their corresponding loading vectors a_q and a_k, where $q, k = 1, \ldots, Q$ with $q \neq k$,
- Each of these loading vectors are penalised with a lasso penalty λ_q that enables variable selection in each data set q, $q = 1, \ldots, Q$,
- Denote C as a $(Q \times Q)$ matrix called the design matrix, which includes the elements $c_{q,k}$. If $c_{q,k} = 1$, then the covariance between the components associated with X_q and X_k is maximised. If $c_{q,k} = 0$ then the covariance is not maximised. Thus, it is possible to constrain the model to only take into account specific pairwise covariances by specifying C (Singh et al., 2019), as illustrated in Figure 13.2. We will come back to the tuning of the design matrix in Section 13.3,
- For a supervised classification analysis (multiblock sPLS-DA) we substitute one X_q with the dummy outcome matrix Y (as detailed in Section 12.2).

For the following dimensions, the X_q matrices are deflated according to Y in a regression framework, as presented in Appendix 10.A, $q = 1, \ldots, Q$. More details about the underlying principles of sGGCA and multiblock sPLS-DA can be found in Appendix 13.A.

Multiblock PLS-DA and multiblock sPLS-DA follow from the theoretical principles of sGCCA, but with further implementations (Singh et al., 2019). In particular, a strong emphasis has been given to graphical outputs (e.g. sample plots, variable plots and circos

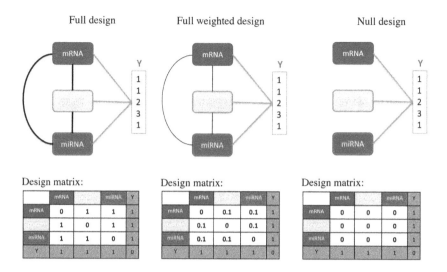

FIGURE 13.2: Example of different design matrices in DIABLO for the multi-omics `breast.TCGA` cancer study illustrated in this chapter. Links or cells in grey are added by default in the function `block.plsda` and `block.splsda` and will not need to be specified. The design matrix is a square numeric matrix of size corresponding to the number of X blocks, with values between 0 and 1. Each value indicates the strength of the relationship to be modelled between two blocks; a value of 0 indicates no relationship, whilst 1 indicates a strong association we wish to model. In a full design (left), the covariance between every possible pair of components associated to each data set is maximised. In a null design (right), only the covariance between each component and the outcome is maximised. In a full weighted design (middle), all data sets are connected but the covariance between pairs of components has a small weight: hence, priority is given to discriminating the outcome, rather than finding correlations between the predictor data sets.

plots, presented in Chapter 6). The input arguments are based on the number of variables to select, rather than the penalisation parameters themselves (that are highly dependent on each data set size). In addition, prediction frameworks that represent a defining feature of these methods have been further developed, and are described below. In Singh et al. (2019), we have drawn particular attention to analyses that both aim to maximise the correlation between data sets and discriminate (and thus predict) the outcome. Our simulation study has shown how the choice of the design matrix can influence one task or the other (as described in Section 13.5).

13.2.2 Prediction in multiblock sPLS-DA

As shown in Figure 13.2, we obtain a component, and therefore a predicted class *per* omics data set during prediction. The predictions are combined by majority vote (the class that has been predicted the most often across all data sets) or by weighted majority vote, where each omics data set weight is defined as the correlation between the latent components associated to that particular data set and those associated with the dummy outcome. The final prediction is the class that obtains the highest vote across all omics data sets. Therefore, the weighted vote gives more importance to the omics data set that is most correlated with the components associated to the outcome. Compared to a majority vote scheme that may lead to discordant but equal votes for a class, the weighted majority vote scheme reduces the number of ties. Ties are indicated as NA in our outputs. We further illustrate the prediction results in Section 13.5.

13.3   Input arguments and tuning

The input is a list of Q data frames (also called *blocks*) $\boldsymbol{X}_1, \ldots, \boldsymbol{X}_Q$ with N rows (the number of samples) and $P_1, \ldots, P_Q$ (the different number of variables in each data frame). In block PLS-DA, $\boldsymbol{y}$ is a factor of length N that indicates the class of each sample. Similar to PLS-DA, $\boldsymbol{y}$ is coded as a dummy matrix $\boldsymbol{Y}$ internally in the function.

The arguments to choose, or tune, include:

- The `design` matrix which indicates which blocks should be connected to maximise the covariance between components, and to what extent. In fact, a compromise needs to be achieved between maximising the correlation between data sets (design value between 0.5 and 1) and maximising the discrimination with the outcome $\boldsymbol{y}$ (design value between 0 and 0.5), as discussed in (Singh et al., 2019). The choice of the design can be based on prior knowledge (*'I expect mRNA and miRNA to be highly correlated'*) or data-driven (e.g. based on a prior analysis, such as a pairwise PLS analysis to examine the correlation between pairs of components associated to each block, as we illustrate in Section 13.5).

- The number of components `ncomp`. The rule of thumb is $K - 1$ where K is the number of classes, similar to PLS-DA. We will use repeated cross-validation, as presented in PLS-DA in Section 12.3, using the `perf()` function.

- The number of variables to select `keepX` for the sparse version, considering a value for each data block and each component. We will use repeated cross-validation using the `tune()` function.

Note:

- *Both **ncomp** and **keepX** could be tuned simultaneously using cross-validation but the problem may quickly become untractable. A more pragmatic approach is to use the framework presented in Section 12.3 where:*
  - *We choose **ncomp** first on a full **block.plsda** model (no variable selection).*
  - *We then tune **keepX** by specifying a grid of values to test per component and per data set. Here we will need to choose the **fold** value in cross-validation, and a grid that is not too thin nor too coarse (often depending on the interpretation potential of the final multi-omics signatures that are obtained).*

13.4   Key outputs

The main outputs of `block.plsda` and `block.splsda` are similar to those from the PLS and PLS-DA approaches. In addition, specific graphical outputs were developed for multi-block integration to support the interpretation of the somewhat complex results (see also Section 6.2).

13.4.1   Graphical outputs

- `plotIndiv()` displays the component scores from each omics data set individually. This plot enables us to visualise the agreement between all data sets at the sample level.

- `plotArrow()` also enables us to assess the similarity, or correlation structure, extracted across data sets (as a superimposition of all sample plots). If there are more than two blocks, the start of all arrows indicates the centroid between all data sets for a given sample, and the tip of each arrow indicates the location of a given sample in each block.

- `plotVar()` represents the correlation circle for all types of variables, either selected by `block.splsda` or above a specified correlation threshold.

- `plotDiablo()` is a matrix scatterplot of the components from each data set for a given dimension; it enables us to check whether the pairwise correlation between two omics has been successfully modelled according to the design.

- `circosPlot()` currently only applies for a `block.splsda` result, and shows pairwise correlations among the variables selected by `block.splsda()` across all data sets. In this plot, variables are represented on the side of the plot and coloured according to their data type. External (optional) lines display their expression levels with respect to each outcome category. This function actually implements an extension of the method used in `plotVar()`, `cim()` and `network()` (see González et al. (2012) and Appendix 6.A).

- `cim()` is similar to the visualisation proposed for PLS-DA and is a simple hierarchical clustering heatmap with samples in columns and selected variables in rows, where we use a specified distance and clustering method (`dist.method` and `clust.method`). A coloured column indicates the type of variables.

- `network()` represents the selected variables in a relevance network graphic.

- `plotLoadings()` shows different panels for each type of variable, and indicates their importance in their respective loading vectors, as well as the sample class they contribute to most.

13.4.2   Numerical outputs

Similar to PLS-DA, we can obtain performance measures of the supervised model, namely:

- The classification performance of the final model using `perf()`,
- The list of variables selected from each data set and associated to each component, and their stability using `selectVar()` and `perf()`, respectively,
- The predicted class of new samples (see Section 13.5),

- The AUC ROC which *complements* the classification performance outputs (as discussed in Section 12.5 and presented in Section 13.5),
- The amount of explained variance per component (but bear in mind that block PLS-DA does not aim to maximise the variance explained, but rather, the covariance with the outcome).

13.5   Case Study: `breast.TCGA`

Human breast cancer is a heterogeneous disease in terms of molecular alterations, cellular composition, and clinical outcome. Breast tumours can be classified into several subtypes, according to their levels of mRNA expression (Sørlie et al., 2001). Here we consider a subset of data generated by The Cancer Genome Atlas Network (Cancer Genome Atlas Network et al., 2012). For the package, data were normalised, and then drastically prefiltered for illustrative purposes.

The data were divided into a *training set* with a subset of 150 samples from the mRNA, miRNA and proteomics data, and a *test set* including 70 samples, but only with mRNA and miRNA data (the proteomics data are missing). The aim of this integrative analysis is to identify a highly correlated multi-omics signature discriminating the breast cancer subtypes Basal, Her2, and LumA.

The `breast.TCGA` (more details can be found in `?breast.TCGA`) is a list containing training and test sets of omics data `data.train` and `data.test` which include:

- `$miRNA`: A data frame with 150 (70) rows and 184 columns in the training (test) data set for the miRNA expression levels,
- `$mRNA`: A data frame with 150 (70) rows and 520 columns in the training (test) data set for the mRNA expression levels,
- `$protein`: A data frame with 150 rows and 142 columns in the training data set for the protein abundance (there are no proteomics in the test set),
- `$subtype`: A factor indicating the breast cancer subtypes in the training (for 150 samples) and test sets (for 70 samples).

This case study covers an interesting scenario where one omic data set is missing in the test set, but because the method generates a set of components per training data set, we can still assess the prediction or performance evaluation using majority or weighted prediction vote (see Section 13.2).

13.5.1   Load the data

To illustrate the multiblock sPLS-DA approach, we will integrate the expression levels of miRNA, mRNA and the abundance of proteins while discriminating the subtypes of breast cancer, then predict the subtypes of the samples in the test set.

The input data is first set up as a list of Q matrices $\boldsymbol{X}_1, \ldots, \boldsymbol{X}_Q$ and a factor indicating the class membership of each sample $\boldsymbol{Y}$. Each data frame in $\boldsymbol{X}$ *should be named* as we will match these names with the `keepX` parameter for the sparse method.

```
library(mixOmics)
data(breast.TCGA)

# Extract training data and name each data frame
# Store as list
X <- list(mRNA = breast.TCGA$data.train$mrna,
          miRNA = breast.TCGA$data.train$mirna,
          protein = breast.TCGA$data.train$protein)

# Outcome
Y <- breast.TCGA$data.train$subtype
summary(Y)


## Basal  Her2  LumA
##    45    30    75
```

13.5.2 Quick start

```
result1.diablo.tcga <- block.plsda(X, Y)      # 1 Run the method
plotIndiv(result1.diablo.tcga)                # 2 Plot the samples
plotVar(result1.diablo.tcga)                  # 3 Plot the variables
```

Here we run `block.plsda` with this minimal code that uses the following default values:

- `ncomp = 2`: the first two sets of components are calculated and are used for graphical outputs,
- `design`: by default a full design is implemented (see Figure 13.2),
- `scale = TRUE`: data are scaled (variance = 1), which is strongly advised here for data integration.

Given the large number of features that are included in the model, the variable plots in `plotVar()` will be difficult to interpret, unless a `cutoff` value is used. To perform variable selection, we use the sparse version `block.splsda` where we arbitrarily specify the argument `keepX` as a list, with list elements names matching the names of the data blocks:

```
# For sparse methods, specify the number of variables to select in each block
# on 2 components with names matching block names
list.keepX <- list(mRNA = c(16, 17), miRNA = c(18,5), protein = c(5, 5))
list.keepX

result.diablo.tcga <- block.splsda(X, Y, keepX = list.keepX) # 1 Run the method
plotIndiv(result.diablo.tcga)                                # 2 Plot the samples
plotVar(result.diablo.tcga)                                  # 3 Plot the variables
```

The default parameters are similar to those from `block.plsda()` with this minimal code. If `keepX` is omitted, then the non-sparse method `block.plsda()` is run by default.

13.5.3 Parameter choice

13.5.3.1 Design matrix

The choice of the design can be motivated by different aspects, including:

- Biological apriori knowledge: Should we expect `mRNA` and `miRNA` to be highly correlated?

- Analytical aims: As further developed in Singh et al. (2019), a compromise needs to be achieved between a classification and prediction task, and extracting the correlation structure of the data sets. A full design with weights = 1 will favour the latter, but at the expense of classification accuracy, whereas a design with small weights will lead to a highly predictive signature. This pertains to the complexity of the analytical task involved as several constraints are included in the optimisation procedure. For example, here we choose a 0.1 weighted model as we are interested in predicting test samples later in this case study.

```
design <- matrix(0.1, ncol = length(X), nrow = length(X),
                dimnames = list(names(X), names(X)))
diag(design) <- 0
design
```

```
##          mRNA miRNA protein
## mRNA      0.0   0.1     0.1
## miRNA     0.1   0.0     0.1
## protein   0.1   0.1     0.0
```

Note however that even with this design, we will still unravel a correlated signature as we require all data sets to explain the same outcome y, as well as maximising pairs of covariances between data sets (Equation (13.1)).

- Data-driven option: we could perform regression analyses with PLS to further understand the correlation between data sets. Here for example, we run PLS with one component and calculate the cross-correlations between components associated to each data set:

```
res1.pls.tcga <- pls(X$mRNA, X$protein, ncomp = 1)
cor(res1.pls.tcga$variates$X, res1.pls.tcga$variates$Y)

res2.pls.tcga <- pls(X$mRNA, X$miRNA, ncomp = 1)
cor(res2.pls.tcga$variates$X, res2.pls.tcga$variates$Y)

res3.pls.tcga <- pls(X$protein, X$miRNA, ncomp = 1)
cor(res3.pls.tcga$variates$X, res3.pls.tcga$variates$Y)
```

```
##          comp1
## comp1   0.9032
##          comp1
## comp1   0.8456
##          comp1
## comp1   0.7982
```

The data sets taken in a pairwise manner are highly correlated, indicating that a design with weights ∼ 0.8–0.9 could be chosen.

13.5.3.2   Number of components

As in the PLS-DA framework presented in Section 12.3, we first fit a `block.plsda` model without variable selection to assess the global performance of the model and choose the number of components to retain. We run `perf()` with 10-fold cross validation repeated ten times for up to five components and with our specified design matrix. Similar to PLS-DA, we obtain the performance of the model with respect to the different prediction distances (Figure 13.3):

```
diablo.tcga <- block.plsda(X, Y, ncomp = 5, design = design)

set.seed(123) # For reproducibility, remove for your analyses
perf.diablo.tcga = perf(diablo.tcga, validation = 'Mfold', folds = 10,
nrepeat = 10)

#perf.diablo.tcga$error.rate   # Lists the different types of error rates

# Plot of the error rates based on weighted vote
plot(perf.diablo.tcga)
```

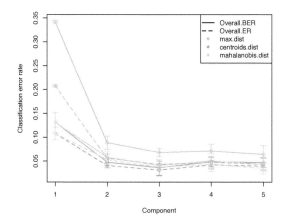

FIGURE 13.3: Choosing the number of components in `block.plsda` using `perf()` with 10× 10-fold CV function in the `breast.TCGA` study. Classification error rates (overall and balanced, see Section 7.3) are represented on the y-axis with respect to the number of components on the x-axis for each prediction distance presented in PLS-DA in Chapter 12 and detailed in Appendix 12.A. Bars show the standard deviation across the ten repeated folds. The plot shows that the error rate reaches a minimum from two to three dimensions.

The performance plot indicates that two components should be sufficient in the final model, and that the centroids distance might lead to better prediction (see details in Appendix 12.A). A balanced error rate (BER) should be considered for further analysis.

The following outputs the optimal number of components according to the prediction distance and type of error rate (overall or balanced), as well as a prediction weighting scheme illustrated further below.

```
perf.diablo.tcga$choice.ncomp$WeightedVote
```

```
##               max.dist centroids.dist mahalanobis.dist
## Overall.ER         3              2                3
## Overall.BER        3              2                3
```

Thus, here we choose our final **ncomp** value:

```
ncomp <- perf.diablo.tcga$choice.ncomp$WeightedVote["Overall.BER", "centroids.dist"]
```

13.5.3.3   Number of variables to select

We then choose the optimal number of variables to select in each data set using the **tune.block.splsda** function. The function **tune()** is run with 10-fold cross validation, but repeated only once (**nrepeat = 1**) for illustrative and computational reasons here. For a thorough tuning process, we advise increasing the **nrepeat** argument to 10–50, or more.

We choose a **keepX** grid that is relatively fine at the start, then coarse. If the data sets are easy to classify, the tuning step may indicate the smallest number of variables to separate the sample groups. Hence, we start our grid at the value **5** to avoid a too small signature that may preclude biological interpretation.

```
# This code may take several min to run, parallelisation is possible
set.seed(123) # For reproducibility with this handbook, remove otherwise
test.keepX <- list(mRNA = c(5:9, seq(10, 25, 5)),
                   miRNA = c(5:9, seq(10, 20, 2)),
                   proteomics = c(seq(5, 25, 5)))

tune.diablo.tcga <- tune.block.splsda(X, Y, ncomp = 2,
                              test.keepX = test.keepX, design = design,
                              validation = 'Mfold', folds = 10, nrepeat = 1,
                              dist = "centroids.dist")
```

Note:

- *For fast computation, we can use parallel computing here – this option is also enabled on a laptop or workstation, see ?tune.block.splsda.*

The number of features to select on each component is returned and stored for the final model:

```
list.keepX <- tune.diablo.tcga$choice.keepX
list.keepX
```

```
## $mRNA
## [1]   8 25
##
```

```
## $miRNA
## [1] 14   5
##
## $protein
## [1] 10   5
```

Note:

- *You can skip any of the tuning steps above, and hard code your chosen **ncomp** and **keepX** parameters.*

13.5.4   Final model

The final multiblock sPLS-DA model includes the tuned parameters and is run as:

```
diablo.tcga <- block.splsda(X, Y, ncomp = ncomp,
                    keepX = list.keepX, design = design)
```

```
## Design matrix has changed to include Y; each block will be
##                linked to Y.
```

```
#diablo.tcga   # Lists the different functions of interest related to that object
```

A warning message informs us that the outcome Y has been included automatically in the design, so that the covariance between each block's component and the outcome is maximised, as shown in the final design output:

```
diablo.tcga$design
```

```
##          mRNA miRNA protein Y
## mRNA      0.0   0.1     0.1 1
## miRNA     0.1   0.0     0.1 1
## protein   0.1   0.1     0.0 1
## Y         1.0   1.0     1.0 0
```

The selected variables can be extracted with the function selectVar(), for example in the mRNA block, along with their loading weights (not output here):

```
# mRNA variables selected on component 1
selectVar(diablo.tcga, block = 'mRNA', comp = 1)
```

Note:

- *The stability of the selected variables can be extracted from the **perf** () function, similar to the example given in the PLS-DA analysis (Section 12.5).*

13.5.5 Sample plots

13.5.5.1 `plotDiablo`

`plotDiablo()` is a diagnostic plot to check whether the correlations between components from each data set were maximised as specified in the design matrix. We specify the dimension to be assessed with the `ncomp` argument (Figure 13.4).

```
plotDiablo(diablo.tcga, ncomp = 1)
```

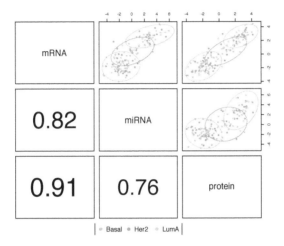

FIGURE 13.4: Diagnostic plot from multiblock sPLS-DA applied on the `breast.TCGA` **study.** Samples are represented based on the specified component (here `ncomp = 1`) for each data set (mRNA, miRNA, and protein). Samples are coloured by breast cancer subtype (Basal, Her2, and LumA) and 95% confidence ellipse plots are represented. The bottom left numbers indicate the correlation coefficients between the first components from each data set. In this example, mRNA expression and protein concentration are highly correlated on the first dimension.

The plot indicates that the first components from all data sets are highly correlated. The colours and ellipses represent the sample subtypes and indicate the discriminative power of each component to separate the different tumour subtypes. Thus, multiblock sPLS-DA is able to extract a strong correlation structure between data sets, as well as discriminate the breast cancer subtypes on the first component.

13.5.5.2 `plotIndiv`

The sample plot with the `plotIndiv()` function projects each sample into the space spanned by the components from *each* block, resulting in a series of graphs corresponding to each data set (Figure 13.5). The optional argument `blocks` can output a specific data set. Ellipse plots are also available (argument `ellipse = TRUE`).

```
plotIndiv(diablo.tcga, ind.names = FALSE, legend = TRUE,
        title = 'TCGA, DIABLO comp 1 - 2')
```

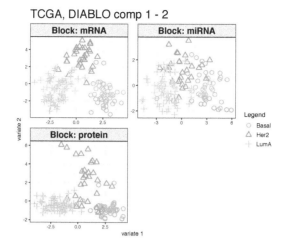

FIGURE 13.5: Sample plot from multiblock sPLS-DA performed on the `breast.TCGA` study. The samples are plotted according to their scores on the first two components for each data set. Samples are coloured by cancer subtype and are classified into three classes: Basal, Her2, and LumA. The plot shows the degree of agreement between the different data sets and the discriminative ability of each data set.

This type of graphic allows us to better understand the information extracted from each data set and its discriminative ability. Here we can see that the LumA group can be difficult to classify in the miRNA data.

Note:

- *Additional variants include the argument* `block = 'average'` *that averages the components from all blocks to produce a single plot. The argument* `block='weighted.average'` *is a weighted average of the components according to their correlation with the components associated with the outcome.*

13.5.5.3 `plotArrow`

In the arrow plot in Figure 13.6, the start of the arrow indicates the centroid between all data sets for a given sample and the tip of the arrow the location of that same sample but in each block. Such graphics highlight the agreement between all data sets at the sample level when modelled with multiblock sPLS-DA.

```
plotArrow(diablo.tcga, ind.names = FALSE, legend = TRUE,
          title = 'TCGA, DIABLO comp 1 - 2')
```

This plot shows that globally, the discrimination of all breast cancer subtypes can be extracted from all data sets, however, there are some dissimilarities at the samples level across data sets (the common information cannot be extracted in the same way across data sets).

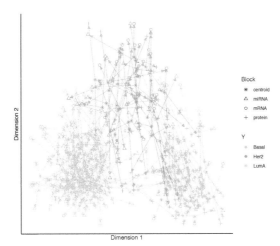

FIGURE 13.6: Arrow plot from multiblock sPLS-DA performed on the `breast.TCGA` study. The samples are projected into the space spanned by the first two components for each data set then overlaid across data sets. The start of the arrow indicates the centroid between all data sets for a given sample and the tip of the arrow the location of the same sample in each block. Arrows further from their centroid indicate some disagreement between the data sets. Samples are coloured by cancer subtype (Basal, Her2, and LumA).

13.5.6 Variable plots

The visualisation of the selected variables is crucial to mine their associations in multiblock sPLS-DA. Here we revisit existing outputs presented in Chapter 6 with further developments for multiple data set integration. All the plots presented provide complementary information for interpreting the results.

13.5.6.1 `plotVar`

The correlation circle plot highlights the contribution of each selected variable to each component. Important variables should be close to the large circle (see Section 6.2). Here, only the variables selected on components 1 and 2 are depicted (across all blocks), see Figure 13.7. Clusters of points indicate a strong correlation between variables. For better visibility we chose to hide the variable names.

```
plotVar(diablo.tcga, var.names = FALSE, style = 'graphics', legend = TRUE,
        pch = c(16, 17, 15), cex = c(2,2,2),
        col = c('darkorchid', 'brown1', 'lightgreen'),
        title = 'TCGA, DIABLO comp 1 - 2')
```

The correlation circle plot shows some positive correlations (between selected miRNA and , between selected and mRNA) and negative correlations between mRNA and miRNA on component 1. The correlation structure is less obvious on component 2, but we observe some key selected features (and miRNA) that seem to highly contribute to component 2.

Note:

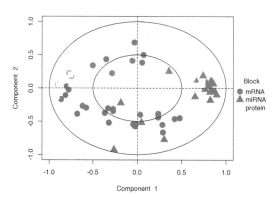

FIGURE 13.7: Correlation circle plot from multiblock sPLS-DA performed on the breast.TCGA study. The variable coordinates are defined according to their correlation with the first and second components for each data set. Variable types are indicated with different symbols and colours, and are overlaid on the same plot. The plot highlights the potential associations within and between different variable types when they are important in defining their own component.

- *These results can be further investigated by showing the variable names on this plot (or extracting their coordinates available from the plot saved into an object, see ?plotVar), and looking at various outputs from selectVar() and plotLoadings().*

- *You can choose to only show specific variable type names, e.g. var.names = c(FALSE, FALSE, TRUE) (where each argument is assigned to a data set in X). Here for example, only the protein names would be output.*

13.5.6.2   circosPlot

The circos plot represents the correlations between variables of different types, represented on the side quadrants. Several display options are possible, to show within and between connections between blocks, and expression levels of each variable according to each class (argument `line = TRUE`). The circos plot is built based on a similarity matrix, which was extended to the case of multiple data sets from González et al. (2012) (see also Section 6.2 and Appendix 6.A). A `cutoff` argument can be further included to visualise correlation coefficients above this threshold in the multi-omics signature (Figure 13.8). The colours for the blocks and correlation lines can be chosen with `color.blocks` and `color.cor`, respectively:

```
circosPlot(diablo.tcga, cutoff = 0.7, line = TRUE,
          color.blocks = c('darkorchid', 'brown1', 'lightgreen'),
          color.cor = c("chocolate3","grey20"), size.labels = 1.5)
```

The circos plot enables us to visualise cross-correlations between data types, and the nature of these correlations (positive or negative). Here we observe that correlations >0.7 are between a few mRNA and some proteins, whereas the majority of strong (negative) correlations are observed between miRNA and mRNA or proteins. The lines indicating the average

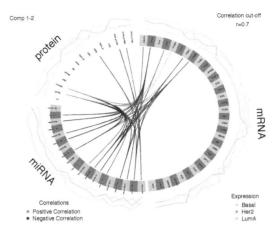

FIGURE 13.8: Circos plot from multiblock sPLS-DA performed on the `breast.TCGA` **study.** The plot represents the correlations greater than 0.7 between variables of different types, represented on the side quadrants. The internal connecting lines show the positive (negative) correlations. The outer lines show the expression levels of each variable in each sample group (Basal, Her2, and LumA).

expression levels per breast cancer subtype indicate that the selected features are able to discriminate the sample groups.

13.5.6.3 `network`

Relevance networks, which are also built on the similarity matrix, can also visualise the correlations between the different types of variables. Each colour represents a type of variable. A threshold can also be set using the argument `cutoff` (Figure 13.9). By default the network includes only variables selected on component 1, unless specified in `comp`.

Note that sometimes the output may not show with Rstudio due to margin issues. We can either use `X11()` to open a new window or save the plot as an image using the arguments `save` and `name.save`, as we show below. An `interactive` argument is also available for the `cutoff` argument, see details in `?network`.

```
# X11()    # Opens a new window
network(diablo.tcga, blocks = c(1,2,3),
        cutoff = 0.4,
        color.node = c('darkorchid', 'brown1', 'lightgreen'),
        # To save the plot, comment out otherwise
        save = 'pdf', name.save = 'diablo-network'
        )
```

The relevance network in Figure 13.9 shows two groups of features of different types. Within each group we observe positive and negative correlations. The visualisation of this plot could be further improved by changing the names of the original features.

Note that the network can be saved in a .gml format to be input into the software Cytoscape, using the R package `igraph` (Csardi et al., 2006):

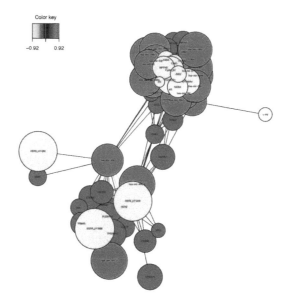

FIGURE 13.9: Relevance network for the variables selected by multiblock sPLS-DA performed on the `breast.TCGA` study on component 1. Each node represents a selected variable with colours indicating their type. The colour of the edges represent positive or negative correlations. Further tweaking of this plot can be obtained, see the help file `?network`.

```
# Not run
library(igraph)
myNetwork <- network(diablo.tcga, blocks = c(1,2,3), cutoff = 0.4)
write.graph(myNetwork$gR, file = "myNetwork.gml", format = "gml")
```

13.5.6.4 plotLoadings

`plotLoadings()` visualises the loading weights of each selected variable on each component and each data set. The colour indicates the class in which the variable has the maximum level of expression (`contrib = 'max'`) or minimum (`contrib = 'min'`), on average (`method = 'mean'`) or using the median (`method = 'median'`), as we show in Figure 13.10.

```
plotLoadings(diablo.tcga, comp = 1, contrib = 'max', method = 'median')
```

The loading plot shows the multi-omics signature selected on component 1, where each panel represents one data type. The importance of each variable is visualised by the length of the bar (i.e. its loading coefficient value). The combination of the sign of the coefficient (positive / negative) and the colours indicate that component 1 discriminates primarily the Basal samples vs. the LumA samples (see the sample plots also). The features selected are highly expressed in one of these two subtypes. One could also plot the second component that discriminates the Her2 samples.

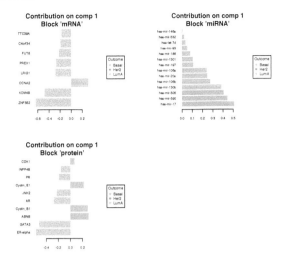

FIGURE 13.10: Loading plot for the variables selected by multiblock sPLS-DA performed on the `breast.TCGA` **study on component 1.** The most important variables (according to the absolute value of their coefficients) are ordered from bottom to top. As this is a supervised analysis, colours indicate the class for which the median expression value is the highest for each feature (variables selected characterise Basal and LumA).

13.5.6.5 cimDiablo

The `cimDiablo()` function is a clustered image map specifically implemented to represent the multi-omics molecular signature expression for each sample. It is very similar to a classical hierarchical clustering (Figure 13.11).

```
cimDiablo(diablo.tcga, color.blocks = c('darkorchid', 'brown1', 'lightgreen'),
          comp = 1, margin=c(8,20), legend.position = "right")
```

According to the CIM, component 1 seems to primarily classify the Basal samples, with a group of overexpressed miRNA and underexpressed mRNA and   . A group of LumA samples can also be identified due to the overexpression of the same mRNA and   . Her2 samples remain quite mixed with the other LumA samples.

13.5.7 Model performance and prediction

We assess the performance of the model using 10-fold cross-validation repeated ten times with the function `perf()`. The method runs a `block.splsda()` model on the pre-specified arguments input from our final object `diablo.tcga` but on cross-validated samples. We then assess the accuracy of the prediction on the left out samples. Since the `tune()` function was used with the `centroid.dist` argument, we examine the outputs of the `perf()` function for that same distance:

```
set.seed(123) # For reproducibility with this handbook, remove otherwise
perf.diablo.tcga <- perf(diablo.tcga,  validation = 'Mfold', folds = 10,
                    nrepeat = 10, dist = 'centroids.dist')
```

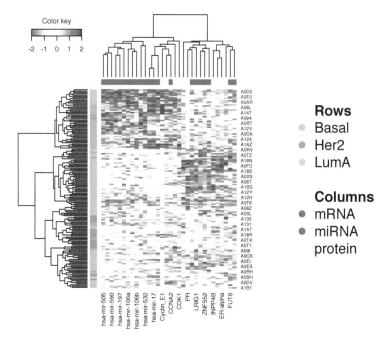

FIGURE 13.11: Clustered Image Map for the variables selected by multiblock sPLS-DA performed on the `breast.TCGA` study on component 1. By default, Euclidean distance and Complete linkage methods are used. The CIM represents samples in rows (indicated by their breast cancer subtype on the left-hand side of the plot) and selected features in columns (indicated by their data type at the top of the plot).

#perf.diablo.tcga   # Lists the different outputs

We can extract the (balanced) classification error rates with `perf.diablo.tcga$error.rate.per.class`, the predicted components associated to Y, or the stability of the selected features with `perf.diablo.tcga$features`.

Here we look at the different performance assessment schemes specific to multiple data set integration, as mentioned in Section 13.2.

First, we output the performance with the majority vote, that is, since the prediction is based on the components associated to their own data set, we can then weight those predictions across data sets according to a majority vote scheme. Based on the predicted classes, we then extract the classification error rate per class and per component:

```
# Performance with Majority vote
perf.diablo.tcga$MajorityVote.error.rate

## $centroids.dist
##              comp1   comp2
## Basal      0.02667 0.04444
## Her2       0.20667 0.12333
## LumA       0.04533 0.00800
```

```
## Overall.ER   0.07200 0.04200
## Overall.BER 0.09289 0.05859
```

The output shows that with the exception of the Basal samples, the classification improves with the addition of the second component.

Another prediction scheme is to weight the classification error rate from each data set according to the correlation between the predicted components and the Y outcome.

```
# Performance with Weighted vote
perf.diablo.tcga$WeightedVote.error.rate
```

```
## $centroids.dist
##                   comp1    comp2
## Basal          0.006667 0.04444
## Her2           0.140000 0.10667
## LumA           0.045333 0.00800
## Overall.ER     0.052667 0.03867
## Overall.BER    0.064000 0.05304
```

Compared to the previous majority vote output, we can see that the classification accuracy is slightly better on component 2 for the subtype Her2.

An AUC plot *per block* is plotted using the function `auroc()`. We have already mentioned in Section 12.5 that the interpretation of this output may not be particularly insightful in relation to the performance evaluation of our methods, but can complement the statistical analysis. For example, here for the miRNA data set once we have reached component 2 (Figure 13.12):

```
auc.diablo.tcga <- auroc(diablo.tcga, roc.block = "miRNA", roc.comp = 2,
                print = FALSE)
```

Figure 13.12 shows that the Her2 subtype is the most difficult to classify with multiblock sPLS-DA compared to the other subtypes.

The `predict()` function associated with a `block.splsda()` object predicts the class of samples from an external test set. In our specific case, one data set is missing in the test set but the method can still be applied. We need to ensure the names of the blocks correspond exactly to those from the training set:

```
# Prepare test set data: here one block (proteins) is missing
data.test.tcga <- list(mRNA = breast.TCGA$data.test$mrna,
                  miRNA = breast.TCGA$data.test$mirna)

predict.diablo.tcga <- predict(diablo.tcga, newdata = data.test.tcga)
# The warning message will inform us that one block is missing

#predict.diablo # List the different outputs
```

The following output is a confusion matrix that compares the real subtypes with the predicted subtypes for a two-component model, for the distance of interest `centroids.dist` and the prediction scheme `WeightedVote`:

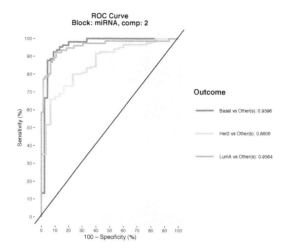

FIGURE 13.12: ROC and AUC based on multiblock sPLS-DA performed on the breast.TCGA study for the miRNA data set after two components. The function calculates the ROC curve and AUC for one class vs. the others. If we set `print = TRUE`, the Wilcoxon test p-value that assesses the differences between the predicted components from one class vs. the others is output.

```
confusion.mat.tcga <- get.confusion_matrix(truth = breast.TCGA$data.test$subtype,
                predicted = predict.diablo.tcga$WeightedVote$centroids.dist[,2])
confusion.mat.tcga
```

```
##          predicted.as.Basal predicted.as.Her2
## Basal                    20                 1
## Her2                      0                13
## LumA                      0                 3
##          predicted.as.LumA
## Basal                    0
## Her2                     1
## LumA                    32
```

From this table, we see that one Basal and one Her2 sample are wrongly predicted as Her2 and Lum A, respectively, and three LumA samples are wrongly predicted as Her2. The balanced prediction error rate can be obtained as:

```
get.BER(confusion.mat.tcga)
```

```
## [1] 0.06825
```

It would be worthwhile at this stage to revisit the chosen design of the multiblock sPLS-DA model to assess the influence of the design on the prediction performance on this test set – even though this back and forth analysis is a biased criterion to choose the design!

13.6 To go further

13.6.1 Additional data transformation for special cases

Similar to the concepts presented in Section 12.6, additional data transformations can be performed on the input data:

- For microbiome or **compositional data** (Section 4.2), the centered log ratio transformation (`logratio = CLR`) can be applied externally to our $N-$integration method using the function `logratio.transfo()`.

- For **repeated measurements** (Section 4.1) we can use the external function `withinVariation()` to extract the within variance matrix, as performed in Lee et al. (2019) with DIABLO.

- A **feature module summarisation**, as presented in Singh et al. (2019) and based on a simple PCA dimension reduction for each gene and metabolite pathway of interest, has also been successfully applied in an asthma study.

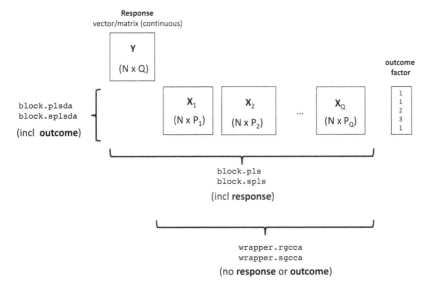

FIGURE 13.13: Schematic view of the multiblock framework for supervised and unsupervised $N-$integration analyses in `mixOmics`. `block.pls` and `block.spls` are regression methods, while `block.plsda` and `block.splsda` are classification methods. `wrapper.rgcca()` and `wrapper.sgcca()` are from the RGCCA package (Tenenhaus and Guillemot, 2017) but with improved input arguments (e.g. `keepX`). The wrapper (and original) functions currently enable an unsupervised analysis.

13.6.2 Other $N-$integration frameworks in `mixOmics`

Several $N-$integration methods are available in `mixOmics`, as represented in Figure 13.13 for various data configurations, and are further detailed in Appendix 13.A:

1. `wrapper.rgcca()` extends rCCA (unsupervised) from the `RGCCA` package by Tenenhaus and Guillemot (2017) to fit into `mixOmics`. It applies in the case where several data sets $X_1, \ldots, X_Q$ are available but there is no response matrix Y. Each data set is deflated symmetrically in a canonical mode fashion (Section 10.2). The function calculates regularisation parameters for each data set, as described in Section 11.3, using the internal shrinkage method (there is no need to specify this parameter as an argument). Both `rCCA()` with the shrinkage method and `wrapper.rgcca()` specifying `tau = 'optimal'` use the same regularisation parameters if we were to integrate two data sets.

2. `wrapper.sgcca()` extends `wrapper.rgcca()` with variable selection. The user needs to specify a list of `keepX` values corresponding to the number of variables to select on each component and each data set. Only the unsupervised framework (canonical mode deflation) is available.

3. `block.pls()` extend sPLS (see Chapter 10) for more than two data sets in a regression framework, and is a special case of `wrapper.sgcca()`. Here we consider the predictor data sets $X_1, \ldots, X_Q$ and a response matrix Y.

4. `block.spls()` extends `block.pls()` for variable selection in the data sets $X_1, \ldots, X_Q$ (input argument `keepX`) and also in the response matrix Y if required (input argument `keepY`) in a regression framework.

For these methods, the usual S3 functions `plotIndiv()`, `plotVar()`, and `selectVar()` apply. However, tuning and performance assessment methods have not yet been developed for these methods.

13.6.3 Supervised classification analyses: concatenation and ensemble methods

Multiple step approaches (Figure 13.13) can include two different types of analysis:

- A concatenation of all data sets prior to applying a classification model.

- Ensemble-based in which a classification model is applied separately to each omics data and the resulting predictions are combined based on average or majority vote (Günther et al., 2012).

However, these approaches can be biased towards certain omics data types – resulting in the over-representation of a specific data type in the variable selection process, and do not account for interactions between omic layers, as we have shown in Singh et al. (2019).

13.6.4 Unsupervised analyses: JIVE and MOFA

Component-based integration methods for multiple data sets have been proposed, including Joint and Individual Variation Explained (JIVE, Lock et al. (2013)) and Multi-Omics Factor Analysis (MOFA, Argelaguet et al. (2018)). The `R.jive` R package has been deprecated, but is still available on GitHub[1] (O'Connell and Lock, 2016). MOFA is available on Bioconductor (R et al., 2020). Comparisons with DIABLO multiblock sPLS-DA were performed in Singh et al. (2019).

---

[1]https://github.com/cran/r.jive

13.7 FAQ

When performing a multi-block analysis, how do I choose my design?

We recommend first relying on some prior biological knowledge you may have on the relationships between data sets. Conduct a few trials on a non-sparse version of `block.plsda()`, and look at the classification performance with `perf()` and `plotDiablo()` before you decide on your final design. See the case study for further explorations.

I have a small number of samples ($N < 10$), should I still tune `keepX` using the `tune()` function?

It is probably not worth it. Try with a few `keepX` values and look at the graphical outputs to see if they make sense. With a small N you can adopt an exploratory approach that does not require a performance assessment.

During `tune()` or `perf()` the code broke down (`system computationally singular`), what happened?

Check that the M value for your M-fold is not too high compared to N (you want $N/M > 6$ as a rule of thumb). Try leave-one-out instead with `validation = 'loo'` and make sure `ncomp` is not too large as you are running on empty matrices after deflation!

My tuning step indicated the selection of only one variable in one of the data sets.

Choose a grid of `keepX` values starting at a higher value (e.g. 5). The algorithm found an optimum with only one variable, either because it is highly discriminatory or because the data are noisy, but it does not stop you from trying for more. In the latter case the first component accounts for 'noise' and the algorithm moves quickly to the second, potentially interesting, dimension.

My Y matrix includes continuous values, what can I do?

You can perform a multi-omics regression with `block.spls()`. We have not yet found a way to tune the results, so you will need to adopt an exploratory approach or support your findings with downstream analyses once you have identified a list of highly correlated features.

Can $N-$integration methods in `mixOmics` fail?

Yes!

- When some (or all) data sets do not explain similar information via linear combinations of variables,
- When some data sets are weakly correlated between each other: you may want to remove those data sets in the integrative analysis,
- When we jump to the integrative analysis without fully understanding the information contained in each data set,
- When data contain a high number of NA values (e.g. methylation): imputation methods such as NIPALS could be envisaged (Section 9.5),
- When there are genotype data (SNP), as these data might need to be managed differently (see Section 4.2.3).

13.8 Additional resources

We are continuing to develop methods and visualisations for $N-$integration, see our website
`http://www.mixomics.org` for our latest updates and applications, for example for single
cell assays.

13.9 Summary

Integrating large-scale molecular omics data sets offers an unprecedented opportunity to
assess molecular interactions at multiple functional levels and provide a more comprehensive
understanding of biological pathways. However, data heterogeneity, platform variability
and a lack of a clear biological question can challenge this kind of data integration. The
multiblock methods in `mixOmics` are designed to address the complex nature of different data
types through dimension reduction, maximisation of the covariance or correlation between
data sets, and feature selection. In this way, we can identify molecular biomarkers across
different functional levels that are correlated and associated with a phenotype of interest.

In particular, multiblock sPLS-DA, or DIABLO, extracts the correlation structure between
omics data sets in a supervised setting, resulting in an improved ability to associate biomarkers
across multiple functional levels whilst classifying samples into known groups, as well as
predict the membership class of new samples. In this case study, we applied DIABLO to a
breast cancer study, integrating three types of omics data sets to identify relevant biomarkers,
while still retaining good classification and predictive performance.

Our statistical integrative framework can benefit a diverse range of research areas with
varying types of study designs, as well as enabling module-based analyses. Importantly,
graphical outputs of our methods assist in the interpretation of such complex analyses and
provide significant biological insights.

Finally, variants of multiblock sPLS-DA are available for different types of data and outcome
matrices.

13.A Appendix: Generalised CCA and variants

We present the underlying mathematical frameworks for the different $N-$integration methods
illustrated in Figure 13.13.

13.A.1 regularised GCCA

Denote Q the normalised, centered and scaled data sets $\boldsymbol{X}_1(N \times P_1)$, $\boldsymbol{X}_2(N \times P_2)$, ...,
$\boldsymbol{X}_Q(N \times P_Q)$ measuring the expression levels of $P_1, \ldots, P_Q$ omics variables on the same N

samples. In rGCCA, the optimisation problem to solve for each dimension $h = 1, \ldots, H$ is (Tenenhaus and Tenenhaus, 2011):

$$\underset{a_1, \ldots, a_Q}{\operatorname{argmax}} \sum_{q,k=1, q \neq k}^{Q} c_{q,k} \, g(\operatorname{cov}(\boldsymbol{X}_q \boldsymbol{a}_q, \boldsymbol{X}_k \boldsymbol{a}_k)) \tag{13.1}$$

$$\text{subject to} \quad \tau_q \|\boldsymbol{a}_q\|^2 + (1 - \tau_q) \operatorname{Var}(\boldsymbol{X}_q \boldsymbol{a}_q) = 1,$$

where $\boldsymbol{a}_q$ are the loading vectors associated with each block $q, q = 1, \ldots, Q$. The function g can be defined as:

- $g(x) = x$ is the Horst scheme, which is used in classical PLS/PLS-DA as well as multiblock sPLS-DA in `mixOmics`.
- $g(x) = |x|$ is the centroid scheme.
- $g(x) = x^2$ is the factorial scheme.

The Horst scheme requires a positive correlation between linear combinations of pairwise data sets, while the centroid and factorial schemes enable a negative correlation between the components. In practice, we have found that the `scheme = 'horst'` gives satisfactory results and therefore it is the default parameter in our methods.

The regularisation parameters `tau` on each block $(\tau_1, \tau_2, \ldots, \tau_Q)$ are internally estimated from the rGCCA method using the shrinkage formula from Schäfer and Strimmer (2005). These parameters enable us to numerically inverse large variance-covariance matrices, as we have seen in Chapter 11.

Notes:

- *The function to run rGCCA is `wrapper.rgcca()` in mixOmics, see `?wrapper.rgcca` for some examples, or the `rgcca()` function from RGCCA (Tenenhaus and Guillemot, 2017).*

- *The matrices $\boldsymbol{X}_1, \ldots, \boldsymbol{X}_Q$ are deflated with a canonical mode, however, other types of deflations can be considered, as mentioned in Tenenhaus et al. (2017), but they have not been implemented in RGCCA so far.*

13.A.2   sparse GCCA

In the same vein as sparse PLS from Lê Cao et al. (2008), sparse GCCA includes lasso penalisations on the loading vectors $\boldsymbol{a}_q$ associated to each data block to perform variable selection. In practice we will specify the arguments `keepX, keepY` in the function `block.spls()` to indicate the number of variables to retain in each block and for each component.

Using the same notations as in Equation (13.1), the optimisation problem to solve is:

$$\underset{a_1, \ldots, a_Q}{\operatorname{argmax}} \sum_{q,k=1, q \neq k}^{Q} c_{q,k} \, g(\operatorname{cov}(\boldsymbol{X}_q \boldsymbol{a}_q, \boldsymbol{X}_k \boldsymbol{a}_k)) \tag{13.2}$$

$$\text{subject to} \quad \|\boldsymbol{a}_q\|_2 = 1 \quad \text{and} \quad \|\boldsymbol{a}_q\|_1 \leq \lambda_q$$

where λ_q is the lasso penalisation on each block $q, q = 1, \ldots, Q$. There is a direct correspondence between the lasso parameter value and the number of variables to select in the arguments `keepX, keepY` in our functions.

Notes:

- *In sparse GCCA, we should not need any regularisation parameters* `tau` *as the model is assumed to be parsimonious (sparse).*

- *The function to run sGCCA is* `wrapper.sgcca()`, *in* `mixOmics`, *see* `?wrapper.sgcca` *for some examples (we implemented the* `keepX` *parameter), or the* `sgcca()` *function from Tenenhaus and Guillemot (2017) (with a lasso penalty parameter).*

13.A.3   sparse multiblock sPLS-DA

DIABLO extends sparse generalised canonical correlation analysis (sGCCA, Tenenhaus et al. (2014)) to a classification or supervised framework. The selection of the correlated molecules across omics levels is performed internally with lasso penalisation on the variable coefficient vectors defining the linear combinations. Since all latent components are scaled in the algorithm, sGCCA maximises the correlation between components but we retain the term *covariance* instead of *correlation* to present the general sGCCA framework.

sGCCA solves the optimisation function for each dimension $h = 1, \ldots, H$:

$$\underset{\boldsymbol{a}_1, \ldots, \boldsymbol{a}_Q}{\operatorname{argmax}} \sum_{q,k=1, q \neq k}^{Q} c_{q,k} \ \operatorname{cov}(\boldsymbol{X}_q \boldsymbol{a}_q, \boldsymbol{X}_k \boldsymbol{a}_k),$$

subject to $||\boldsymbol{a}_q||_2 = 1$ and $||\boldsymbol{a}_q||_1 \leq \lambda_q$ for all $1 \leq q \leq Q$,

where $\boldsymbol{a}_q$ is the variable coefficient or loading vector for a given dimension h associated to the current residual matrix of the data set $\boldsymbol{X}_q$, $q = 1, \ldots, Q$. $\boldsymbol{C} = \{c_{q,k}\}_{q,k}$ is a $(Q \times Q)$ design matrix that specifies whether data sets should be connected, as we described in Section 13.3, with $q, k = 1, \ldots, Q$, $q \neq k$. Similar to sGCCA, Equation (13.2), λ_q is a non-negative parameter that controls the amount of shrinkage and thus the number of non-zero coefficients in $\boldsymbol{a}_q$ for a given dimension h for data set $\boldsymbol{X}_q$. The penalisation enables the selection of a subset of variables with non-zero coefficients that define each component score $\boldsymbol{t}_q = \boldsymbol{X}_q \boldsymbol{a}_q$ for a given dimension h. The result is the identification of sets of variables that are highly correlated *between* and *within* omics data sets. In multiblock sPLS-DA, we use the Horst scheme.

The difference with sGCCA (Equation (13.2)) is the regression framework. Denote $\{\boldsymbol{a}_1^1, \ldots \boldsymbol{a}_Q^1\}$ the first set of coefficient vectors for dimension $h = 1$, obtained by maximising Equation (13.3) with $\boldsymbol{X}_q^1 = \boldsymbol{X}_q$. For $h = 2$, we maximise Equation (13.3) using the residual matrices $\boldsymbol{X}_q^2 = \boldsymbol{X}_q^1 - \boldsymbol{t}_q^1 \boldsymbol{a}_q^1$, for each $q = 1, \ldots, Q$, *except* for the dummy $\boldsymbol{Y}^2$ outcome that is deflated with respect to the average of the components $\{\boldsymbol{t}_1^1, \ldots, \boldsymbol{t}_Q^1\}$ (refer to how we calculate the regression coefficient vector $\boldsymbol{d}$ associated with $\boldsymbol{Y}$ in the PLS algorithm in Appendix 10.A) and so on for the other dimensions. This process is repeated until a sufficient number of dimensions H (or set of components) is obtained. The underlying assumption in multiblock sPLS-DA is that the major source of common biological variation can be extracted via the sets of component scores $\{\boldsymbol{t}_1^h, \ldots, \boldsymbol{t}_Q^h\}$, $h = 1, \ldots, H$, while any unwanted variation due to heterogeneity across the data sets $\boldsymbol{X}_q$ does not impact the statistical model. The optimisation problem is solved using a monotonically convergent algorithm (Tenenhaus et al., 2014). In our implementation, the number of dimensions is the same for each block.

14

P−data integration

14.1 Why use P−integration methods?

Biological findings from high-throughput experiments often suffer from poor reproducibility as the studies include a small number of samples and a large number of variables. As a consequence, the identified gene signatures across studies can be vastly different. The integration of independent data sets measured on the same common P features is a useful approach to increase sample size and gain statistical power (Lazzeroni and Ray, 2012). In this context, the challenge is to accommodate for systematic differences that arise due to differences between protocols, geographical sites or the use of different technological platforms for the same type of omics data. Systematic unwanted variation, also called batch effects, often acts as a strong confounder in the statistical analysis and may lead to spurious results and conclusions if not accounted for in the statistical analysis (see Section 2.2).

In this chapter we introduce MINT (Multivariate INTegration, Rohart et al. (2017b)), a method based on multi-group PLS that includes information about samples belonging to independent groups or studies (Eslami et al., 2014). We focus on a supervised classification analysis where the aim is to predict the class of new samples from external studies whilst identifying a 'universal' signature across studies. Most of the concepts presented in Chapter 12 will apply here. A regression framework is also available but still represents work in progress.

14.1.1 Biological question

I want to analyse different studies related to the same biological question, using similar omics technologies. Can I combine the data sets while accounting for the variation between studies? Can I discriminate the samples based on their outcome category? Which variables are discriminative across all studies? Can they constitute a signature that predicts the class of external samples?

14.1.2 Statistical point of view

We refer to this framework as P−integration as every data set is acquired on the same P molecules, under similar conditions (introduced in Section 4.3). In a classification framework, we consider an outcome variable such as treatment, where all treatments are represented within each study. In a regression framework, we consider a response matrix in each study (Figure 14.1). We account for the systematic platform variation within the model via *sparse multi-group PLS* models, where we specify the membership of each sample within each study, whilst maximising the covariance with the outcome or the response $\boldsymbol{Y}$. The result

DOI: 10.1201/9781003026860-14

is the identification of a robust molecular signature across multiple studies to discriminate biological conditions.

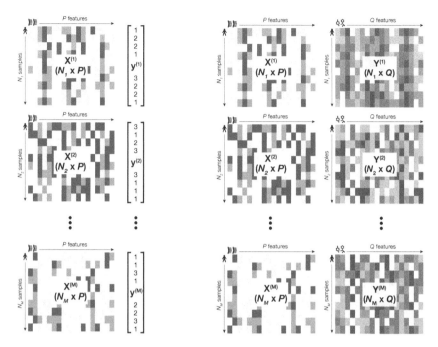

FIGURE 14.1: Two frameworks for $P−$integration in `mixOmics`. Left-hand side: we consider M predictor data sets $\boldsymbol{X}^{(1)}, \ldots, \boldsymbol{X}^{(M)}$ and their respective outcomes $\boldsymbol{y}^{(1)}, \ldots, \boldsymbol{y}^{(M)}$ in a classification framework analysed with *sparse multi-group PLS-DA*. Right-hand side: we consider M predictor data sets $\boldsymbol{X}^{(1)}, \ldots, \boldsymbol{X}^{(M)}$ and their respective response matrices $\boldsymbol{Y}^{(1)}, \ldots, \boldsymbol{Y}^{(M)}$ in a regression framework analysed with *sparse multi-group PLS*.

14.2 Principle

14.2.1 Motivation

$P−$integration (sometimes called 'horizontal' integration, e.g. in Richardson et al. (2016)) is commonly approached in a sequential manner: A first step consists of removing batch effects using, for example, ComBat (Johnson et al., 2007), FAbatch (Hornung et al., 2016), Batch Mean-Centering (Sims et al., 2008), LMM-EH-PS (Listgarten et al., 2010), RUV-2 (Gagnon-Bartsch and Speed, 2012), or YuGene (Lê Cao et al., 2014). A second step then fits a statistical model on the batch-corrected data. A range of classification methods can be applied if we are interested in classifying biological samples and predicting the class membership of new samples, for example with machine learning approaches (e.g. random forests or multivariate linear approaches such as LDA or PLS-DA.

The major pitfall of the sequential approach is a risk of overfitting the training set, as discussed in Section 1.4, leading to non-reproducible signatures on the test set or external studies. MINT mitigates these problems by including a study or group structure in a PLS-based model to avoid sequential steps or extensive subsampling (Rohart et al., 2017b).

In this chapter, we will use the following notations: we consider M data sets denoted $\boldsymbol{X}^{(1)}(N_1 \times P)$, $\boldsymbol{X}^{(2)}(N_2 \times P)$, ..., $\boldsymbol{X}^{(M)}(N_M \times P)$ measured on the same P predictors but from independent studies, with $N = \sum_{m=1}^{M} N_m$. Each data set $\boldsymbol{X}^{(m)}$ has an associated *dummy* indicator outcome $\boldsymbol{Y}^{(m)}$ in which all K classes are represented, $m = 1, \ldots, M$. We denote $\boldsymbol{X}$ $(N \times P)$ and $\boldsymbol{Y}$ $(N \times K)$ the concatenation of all $\boldsymbol{X}^{(m)}$ and $\boldsymbol{Y}^{(m)}$, respectively. Each variable from the data set $\boldsymbol{X}^{(m)}$ and $\boldsymbol{Y}^{(m)}$ is centered and has unit variance internally in the method.

Note:

- *If an internal known batch effect is present in a study, this study should be split according to that batch effect factor into several sub-studies that are considered as independent.*

14.2.2 Multi-group sPLS-DA

In a classification framework, we seek for a common projection space for all studies defined on a small subset of variables that discriminate the outcome of interest (Figure 14.2). The identified variables share common information and therefore aim to represent a reproducible signature across all studies. For each component h, we solve:

$$\max_{\boldsymbol{a}, \boldsymbol{b}} \sum_{m=1}^{M} N_m \ \mathrm{cov}(\boldsymbol{X}^{(m)}\boldsymbol{a}, \boldsymbol{Y}^{(m)}\boldsymbol{b}), \tag{14.1}$$
$$\text{subject to} \quad ||\boldsymbol{a}||_2 = 1 \text{ and } ||\boldsymbol{a}||_1 \leq \lambda$$

where $\boldsymbol{a}$ and $\boldsymbol{b}$ are the *global loading vectors* common to all studies. The *partial PLS-components* $\boldsymbol{t}^{(m)} = \boldsymbol{X}^{(m)}\boldsymbol{a}$ and $\boldsymbol{u}^{(m)} = \boldsymbol{Y}^{(m)}\boldsymbol{b}$ are study-specific, while the *global PLS-components* $\boldsymbol{t} = \boldsymbol{X}\boldsymbol{a}$ and $\boldsymbol{u} = \boldsymbol{Y}\boldsymbol{b}$ represent the global space where all samples are projected. Partial loading vectors are calculated based on a local regression on the variables from either $\boldsymbol{X}^{(m)}$ or $\boldsymbol{Y}^{(m)}$ onto the partial component $\boldsymbol{u}^{(m)}$ and $\boldsymbol{t}^{(m)}$, respectively, similar to the PLS presented in Appendix 10.A (see also Rohart et al. (2017b)), $m = 1, \ldots, M$.

Residual (deflated) matrices are calculated for each iteration of the algorithm based on the global components and loading vectors. Thus MINT models the study structure during the integration process. The penalisation parameter λ controls the amount of shrinkage and thus the number of non-zero weights in the global loading vector $\boldsymbol{a}$.

Equation (14.1) is an extension of multi-group PLS from Eslami et al. (2014) for classification analysis and feature selection. Similarly to sPLS-DA introduced in Chapter 12, MINT selects a combination of features on each PLS-component. In addition, MINT was also extended to an integrative regression framework, but the accompanying tuning and performance functions are still in development.

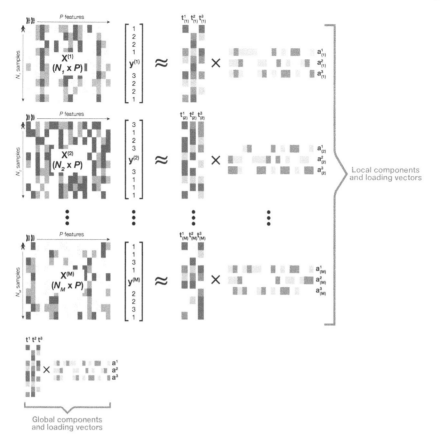

FIGURE 14.2: Multi-group PLS-DA decomposition with MINT. The predictor
M matrices $\boldsymbol{X}^{(1)}, \ldots, \boldsymbol{X}^{(M)}$ and their respective outcomes $\boldsymbol{y}^{(1)}, \ldots, \boldsymbol{y}^{(M)}$ are decomposed
into a set of global loading vectors and loading components across all studies, and a set of
partial loading vectors and components for each study. Similar to PLS-DA, we maximise
the covariance between the $\boldsymbol{X}$ data sets and the $\boldsymbol{Y}$ outcomes (internally transformed into
dummy matrices).

14.3 Input arguments and tuning

Since MINT is based on PLS-DA, we use a similar framework as presented in Section 12.3 to
tune the number of components and variables to select. Caution needs to be applied when
we input the data.

14.3.1 Data input checks

When applying MINT, some considerations need be taken into account:

- To reduce unwanted systematic variation between studies, the method centers and scales
  each study as an initial step. Therefore, we can only include studies with a sample size
  greater than 3. In addition, all outcome categories need to be represented in each study.

- Sample annotation is important in this analysis: some research groups or labs may define the class membership of the samples differently. In the `stemcells` example, the biological definition of hiPSC differs across research groups. Thus, the expertise and exhaustive screening required to homogeneously annotate samples may hinder $P-$integration if not considered with caution.

- When possible, we advise having access to raw data sets from each study to homogenise normalisation techniques as much as possible. Variation in the normalisation processes of different data sets produces unwanted variation between studies.

14.3.2 Number of components

In MINT we take advantage of the independence between studies to evaluate the performance based on a CV technique called 'Leave-One-Group-Out Cross-Validation' (LOGOCV, Rohart et al. (2017b)). LOGOCV performs CV where each study is left out once, and is computationally very efficient. The aim is to reflect a realistic prediction of independent external studies. We can use the function `perf()` and subsequently choose the prediction distance for the `keepX` tuning step described below.

Note:

- *LOGOCV does not need to be repeated as the partitioning is not random.*

14.3.3 Number of variables to select per component

Similar to sPLS-DA, we use cross-validation, but this time based on LOGOCV to choose the number of variables to select in the `tune()` function.

14.4 Key outputs

14.4.1 Graphical outputs

The graphical outputs presented in PLS-DA are available for MINT (see Section 12.4). In addition, the $P-$integration framework takes advantage of the partial and global components from the multi-group PLS approach:

- The set of partial components $\{t^h_{(1)}, \ldots, t^h_{(M)}\}$ provides outputs specific to each study in `plotIndiv()`, where $h = 1, \ldots, H$ is the dimension. The sample plots enable a quality control step to identify studies that may cluster outcome classes differently to other studies (i.e. potential 'outlier' studies).

- The function `plotLoadings()` displays the loading coefficients of the features globally selected by the model but represented individually in each study to visualise potential discrepancies between studies. Visualisation of the global loading vectors is also possible (argument `study = 'all.partial'` or `'global'`).

14.4.2   Numerical outputs

The outputs are similar to those in sPLS-DA introduced in Section 12.4, including:

- The classification performance of the final model (using `perf()`).
- The stability of the selected variables (using `selectVar()` and `perf()`).
- The predicted class of new samples (detailed in Section 12.5).
- The AUC ROC which *complements* the classification performance outputs (see Section 14.5).

14.5   Case Study: `stemcells`

We integrate four transcriptomics studies of microarray stem cells (125 samples in total). The original data set from the Stemformatics database[1] (Wells et al., 2013) was reduced to fit into the package, and includes a randomly-chosen subset of the expression levels of 400 genes. The aim is to classify three types of human cells: human fibroblasts (Fib) and human induced Pluripotent Stem Cells (hiPSC & hESC).

There is a biological hierarchy among the three cell types. On the one hand, differences between pluripotent (hiPSC and hESC) and non-pluripotent cells (Fib) are well-characterised and are expected to contribute to the main biological variation. On the other hand, hiPSC are genetically reprogrammed to behave like hESC and both cell types are commonly assumed to be alike. However, differences have been reported in the literature (Chin et al. (2009), Newman and Cooper (2010)). We illustrate the use of MINT to address sub-classification problems in a single analysis.

14.5.1   Load the data

We first load the data from the package and set up the categorical outcome Y and the `study` membership:

```
library(mixOmics)
data(stemcells)

# The combined data set X
X <- stemcells$gene
dim(X)

## [1] 125 400

# The outcome vector Y:
Y <- stemcells$celltype
length(Y)
```

---

[1] www.stemformatics.org

```
## [1] 125
```

```
summary(Y)
```

```
## Fibroblast        hESC        hiPSC
##          30          37          58
```

We then store the vector indicating the sample membership of each independent study:

```
study <- stemcells$study
```

```
# Number of samples per study:
summary(study)
```

```
##  1  2  3  4
## 38 51 21 15
```

```
# Experimental design
table(Y,study)
```

```
##                study
## Y                1  2  3  4
##    Fibroblast    6 18  3  3
##    hESC         20  3  8  6
##    hiPSC        12 30 10  6
```

14.5.2 Quick start

14.5.2.1 MINT PLS-DA

We need to specify the argument `study`:

```
res.mint.plsda.stem <- mint.plsda(X, Y, study = study)    # 1 Run the method
plotIndiv(res.mint.plsda.stem)                            # 2 Plot the samples
plotVar(res.mint.plsda.stem)                              # 3 Plot the variables
```

We look into `?mint.plsda` for the default arguments in this function:

- `ncomp = 2`: The first two multi-group PLS-DA components are calculated.
- `scale = TRUE`: Each study is centered and scaled (each variable has a mean of 0 and variance of 1).

The sample plot `plotIndiv()` will automatically colour the samples according to their class membership (retrieved internally from the input Y). The default visualisation is to show the global components (use the argument `study = "all.partial"` otherwise).

The correlation circle plot `plotVar()` appears very cluttered, hence there will be some benefit in using the sparse version of multi-group PLS-DA.

14.5.2.2   MINT sPLS-DA

In this example, we arbitrarily specify a specific number of variables to select on each dimension, here 10 and 5 for example.

```
mint.splsda.result <- mint.splsda(X, Y, study = study,
                          keepX = c(10,5))   # 1 Run the method
plotIndiv(mint.splsda.result)               # 2 Plot the samples
plotVar(mint.splsda.result)                 # 3 Plot the variables
selectVar(mint.splsda.result, comp = 1)$name    # Selected variables on comp 1
plotLoadings(mint.splsda.result, method = 'mean', contrib = 'max')
```

The default parameters in `?mint.splsda` are similar to `mint.plsda` presented above, with the addition of:

- `keepX`: set to P (i.e. select all variables) if the argument is not specified.

The selected variables are ranked by absolute coefficient weight from each loading vector in `selectVar()`. The class the selected variables describe (according to the mean/median min or max expression values) is represented in `plotLoadings()`. The argument in `plotLoadings()` can be changed to `study = "all.partial"` to visualise the partial loading vectors in each study.

14.5.3   Example: MINT PLS-DA

We first perform a MINT PLS-DA with all variables included in the model and `ncomp = 5` components. The `perf()` function is used to estimate the performance of the model using LOGOCV, and to choose the optimal number of components for our final model (see Figure 14.3).

```
mint.plsda.stem <- mint.plsda(X = X, Y = Y, study = study, ncomp = 5)

set.seed(2543) # For reproducible results here, remove for your own analyses
perf.mint.plsda.stem <- perf(mint.plsda.stem)

plot(perf.mint.plsda.stem)
```

Based on the performance plot (Figure 14.3), `ncomp = 2` seems to achieve the best performance for the centroid distance, and `ncomp = 1` for the Mahalanobis distance in terms of BER. Additional numerical outputs such as the BER and overall error rates per component, and the error rates per class and per prediction distance, can be output:

```
perf.mint.plsda.stem$global.error$BER
# Type also:
# perf.mint.plsda.stem$global.error


##         max.dist centroids.dist mahalanobis.dist
## comp1    0.5189       0.3504          0.3504
## comp2    0.3739       0.3371          0.3798
```

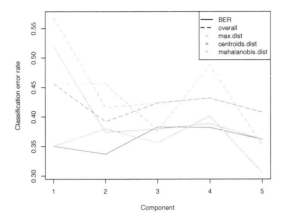

FIGURE 14.3: Choosing the number of components in `mint.plsda` **using** `perf()` **with LOGOCV in the** `stemcells` **study.** Classification error rates (overall and balanced, see Section 7.3) are represented on the y-axis with respect to the number of components on the x-axis for each prediction distance (see Section 12.2 and Appendix 12.A for a refresher). The plot shows that the error rate reaches a minimum from one component with the BER and centroids distance.

```
## comp3    0.3796        0.3829        0.3559
## comp4    0.3886        0.3821        0.4016
## comp5    0.3616        0.3616        0.3051
```

While we may want to focus our interpretation on the first component, we run a final MINT PLS-DA model for `ncomp = 2` to obtain 2D graphical outputs (Figure 14.4):

```
final.mint.plsda.stem <- mint.plsda(X = X, Y = Y, study = study, ncomp = 2)

#final.mint.plsda.stem # Lists the different functions

plotIndiv(final.mint.plsda.stem, legend = TRUE, title = 'MINT PLS-DA',
          subtitle = 'stem cell study', ellipse = T)
```

The sample plot (Figure 14.4) shows that fibroblasts are separated on the first component. We observe that while deemed not crucial for an optimal discrimination, the second component seems to help separate hESC and hiPSC further. The effect of study after MINT modelling is not strong.

We can compare this output to a classical PLS-DA to visualise the study effect (Figure 14.5):

```
plsda.stem <- plsda(X = X, Y = Y, ncomp = 2)

plotIndiv(plsda.stem, pch = study,
          legend = TRUE, title = 'Classic PLS-DA',
          legend.title = 'Cell type', legend.title.pch = 'Study')
```

MINT PLS-DA

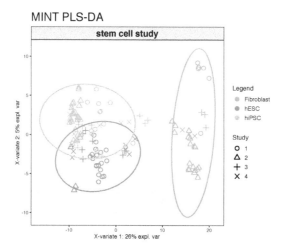

FIGURE 14.4: Sample plot from the MINT PLS-DA performed on the `stemcells` gene expression data. Samples are projected into the space spanned by the first two components. Samples are coloured by their cell types and symbols indicate the study membership. Component 1 discriminates fibroblast vs. the others, while component 2 discriminates some of the hiPSC vs. hESC.

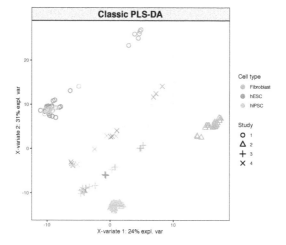

FIGURE 14.5: Sample plot from a classic PLS-DA performed on the `stemcells` gene expression data that highlights the study effect (indicated by symbols). Samples are projected into the space spanned by the first two components. We still do observe some discrimination between the cell types.

14.5.4 Example: MINT sPLS-DA

The MINT PLS-DA model shown earlier is built on all 400 genes in $\boldsymbol{X}$, many of which may be uninformative to characterise the different classes. Here we aim to identify a small subset of genes that best discriminate the classes.

14.5.4.1 Number of variables to select

We can choose the `keepX` parameter using the `tune()` function for a MINT object. The function performs LOGOCV for different values of `test.keepX` provided on each component, and no repeat argument is needed. Based on the mean classification error rate (overall error rate or BER) and a centroids distance, we output the optimal number of variables `keepX` to be included in the final model.

```
set.seed(2543)  # For a reproducible result here, remove for your own analyses
tune.mint.splsda.stem <- tune(X = X, Y = Y, study = study,
                ncomp = 2, test.keepX = seq(1, 100, 1),
                method = 'mint.splsda', #Specify the method
                measure = 'BER',
                dist = "centroids.dist")

#tune.mint.splsda.stem # Lists the different types of outputs

# Mean error rate per component and per tested keepX value:
#tune.mint.splsda.stem$error.rate[1:5,]
```

The optimal number of variables to select on each specified component:

```
tune.mint.splsda.stem$choice.keepX
```

```
## comp1 comp2
##    24    45
```

```
plot(tune.mint.splsda.stem)
```

The tuning plot in Figure 14.6 indicates the optimal number of variables to select on component 1 (24) and on component 2 (45). In fact, whilst the BER decreases with the addition of component 2, the standard deviation remains large, and thus only one component is optimal. However, the addition of this second component is useful for the graphical outputs, and also to attempt to discriminate the hESC and hiPCS cell types.

Note:

- *As shown in the quick start example, the tuning step can be omitted if you prefer to set arbitrary* `keepX` *values.*

14.5.4.2 Final MINT sPLS-DA model

Following the tuning results, our final model is as follows (we still choose a model with two components in order to obtain 2D graphics):

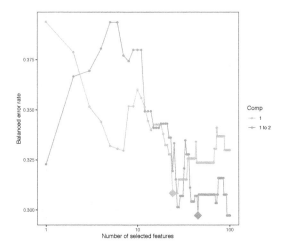

FIGURE 14.6: Tuning `keepX` in MINT sPLS-DA performed on the `stemcells` gene expression data. Each coloured line represents the balanced error rate (y-axis) per component across all tested `keepX` values (x-axis). The diamond indicates the optimal `keepX` value on a particular component which achieves the lowest classification error rate as determined with a one-sided $t-$test across the studies.

```
final.mint.splsda.stem <- mint.splsda(X = X, Y = Y, study = study, ncomp = 2,
                        keepX = tune.mint.splsda.stem$choice.keepX)

#mint.splsda.stem.final # Lists useful functions that can be used with a MINT object
```

14.5.4.3   Sample plots

The samples can be projected on the global components or alternatively using the partial components from each study (Figure 14.7).

```
plotIndiv(final.mint.splsda.stem, study = 'global', legend = TRUE,
        title = 'Stem cells, MINT sPLS-DA',
        subtitle = 'Global', ellipse = T)


plotIndiv(final.mint.splsda.stem, study = 'all.partial', legend = TRUE,
        title = 'Stem cells, MINT sPLS-DA',
        subtitle = paste("Study",1:4))
```

The visualisation of the partial components enables us to examine each study individually and check that the model is able to extract a good agreement between studies.

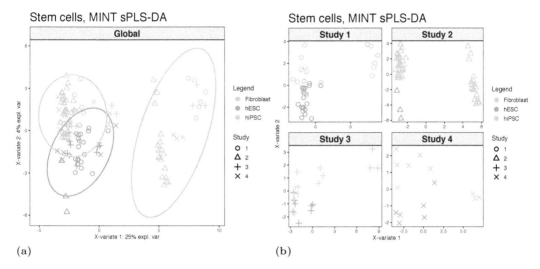

(a)   (b)

FIGURE 14.7: Sample plots from the MINT sPLS-DA performed on the `stemcells` **gene expression data.** Samples are projected into the space spanned by the first two components. Samples are coloured by their cell types and symbols indicate study membership. (a) Global components from the model with 95% ellipse confidence intervals around each sample class. (b) Partial components per study show a good agreement across studies. Component 1 discriminates fibroblasts vs. the rest, component 2 discriminates further hESC vs. hiPSC.

14.5.4.4   Variable plots

Correlation circle plot

We can examine our molecular signature selected with MINT sPLS-DA. The correlation circle plot, presented in Section 6.2, highlights the contribution of each selected transcript to each component (close to the large circle), and their correlation (clusters of variables) in Figure 14.8.

```
plotVar(final.mint.splsda.stem)
```

We observe a subset of genes that are strongly correlated and negatively associated to component 1 (negative values on the x-axis), which are likely to characterise the groups of samples hiPSC and hESC, and a subset of genes positively associated to component 1 that may characterise the fibroblast samples (and are negatively correlated to the previous group of genes).

Note:

- *We can use the* ***var.name*** *argument to show gene name ID, as shown in Section 12.5 for PLS-DA.*

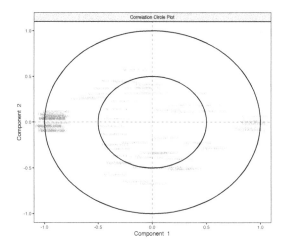

FIGURE 14.8: Correlation circle plot representing the genes selected by MINT sPLS-DA performed on the `stemcells` gene expression data to examine the association of the genes selected on the first two components. We mainly observe two groups of genes, either positively or negatively associated with component 1 along the x-axis. This graphic should be interpreted in conjunction with the sample plot.

Clustered Image Maps

The Clustered Image Map represents the expression levels of the gene signature per sample, similar to a PLS-DA object (see Section 12.5). Here we use the default Euclidean distance and Complete linkage in Figure 14.9 for a specific component (component 1).

```
# If facing margin issues, use either X11() or save the plot using the
# arguments save and name.save
cim(final.mint.splsda.stem, comp = 1, margins=c(10,5),
    row.sideColors = color.mixo(as.numeric(Y)), row.names = FALSE,
    title = "MINT sPLS-DA, component 1")
```

As expected and observed from the sample plot in Figure 14.7, we observe in the CIM that the expression of the genes selected on component 1 discriminates primarily the fibroblasts vs. the other cell types.

Relevance networks

Relevance networks can also be plotted for a PLS-DA object, but would only show the association between the selected genes and the cell type (dummy variable in Y as an outcome category) as shown in Figure 14.10. Only the variables selected on component 1 are shown (`comp = 1`):

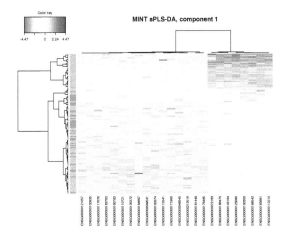

FIGURE 14.9: Clustered Image Map of the genes selected by MINT sPLS-DA on the stemcells gene expression data for component 1 only. A hierarchical clustering based on the gene expression levels of the selected genes on component 1, with samples in rows coloured according to cell type showing a separation of the fibroblasts vs. the other cell types.

```
# If facing margin issues, use either X11() or save the plot using the
# arguments save and name.save
network(final.mint.splsda.stem, comp = 1,
        color.node = c(color.mixo(1), color.mixo(2)),
        shape.node = c("rectangle", "circle"))
```

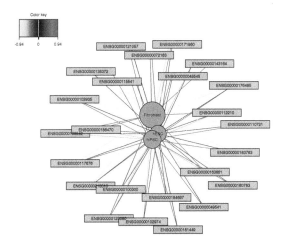

FIGURE 14.10: Relevance network of the genes selected by MINT sPLS-DA performed on the stemcells gene expression data for component 1 only. Associations between variables from X and the dummy matrix Y are calculated as detailed in Appendix 6.A. Edges indicate high or low associations between the genes and the different cell types.

Variable selection and loading plots

The `selectVar()` function outputs the selected transcripts on the first component along with their loading weight values. We consider variables as important in the model when their absolute loading weight value is high. In addition to this output, we can compare the stability of the selected features across studies using the `perf()` function, as shown in PLS-DA in Section 12.5.

```
# Just a head
head(selectVar(final.mint.plsda.stem, comp = 1)$value)
```

```
##                     value.var
## ENSG00000181449    -0.09764
## ENSG00000123080     0.09606
## ENSG00000110721    -0.09595
## ENSG00000176485    -0.09457
## ENSG00000184697    -0.09387
## ENSG00000102935    -0.09370
```

The `plotLoadings()` function displays the coefficient weight of each selected variable in each study and shows the agreement of the gene signature across studies (Figure 14.11). Colours indicate the class in which the mean expression value of each selected gene is maximal. For component 1, we obtain:

```
plotLoadings(final.mint.splsda.stem, contrib = "max", method = 'mean', comp=1,
             study="all.partial", title="Contribution on comp 1",
             subtitle = paste("Study",1:4))
```

Several genes are consistently over-expressed on average in the fibroblast samples in each of the studies, however, we observe a less consistent pattern for the other genes that characterise hiPSC and hESC. This can be explained as the discrimination between both classes is challenging on component 1 (see sample plot in Figure 14.7).

14.5.4.5 Classification performance

We assess the performance of the MINT sPLS-DA model with the `perf()` function. Since the previous tuning was conducted with the distance `centroids.dist`, the same distance is used to assess the performance of the final model. We do not need to specify the argument `nrepeat` as we use LOGOCV in the function.

```
set.seed(123)  # For reproducible results here, remove for your own study
perf.mint.splsda.stem.final <- perf(final.mint.plsda.stem, dist = 'centroids.dist')
```

```
perf.mint.splsda.stem.final$global.error
```

```
## $BER
##         centroids.dist
## comp1          0.3504
## comp2          0.3371
```

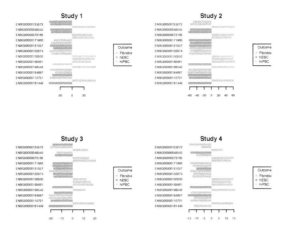

FIGURE 14.11: Loading plots of the genes selected by the MINT sPLS-DA performed on the `stemcells` data, on component 1 per study. Each plot represents one study, and the variables are coloured according to the cell type they are maximally expressed in, on average. The length of the bars indicate the loading coefficient values that define the component. Several genes distinguish between fibroblasts and the other cell types, and are consistently overexpressed in these samples across all studies. We observe slightly more variability in whether the expression levels of the other genes are more indicative of hiPSC or hESC cell types.

```
## 
## $overall
##         centroids.dist
## comp1          0.456
## comp2          0.392
## 
## $error.rate.class
## $error.rate.class$centroids.dist
##               comp1  comp2
## Fibroblast 0.0000 0.0000
## hESC       0.1892 0.4595
## hiPSC      0.8621 0.5517
```

The classification error rate per class is particularly insightful to understand which cell types are difficult to classify, hESC and hiPS – whose mixture can be explained for biological reasons.

14.5.5   Take a detour

14.5.5.1   AUC

An AUC plot for the integrated data can be obtained using the function `auroc()` (Figure 14.12).

Remember that the AUC incorporates measures of sensitivity and specificity for every possible cut-off of the predicted dummy variables. However, our PLS-based models rely on prediction distances, which can be seen as a determined optimal cut-off. Therefore, the ROC and AUC criteria may not be particularly insightful in relation to the performance evaluation of our supervised multivariate methods, but can complement the statistical analysis (from Rohart et al. (2017a)).

```
auroc(final.mint.splsda.stem, roc.comp = 1)
```

We can also obtain an AUC plot per study by specifying the argument `roc.study`:

```
auroc(final.mint.splsda.stem, roc.comp = 1, roc.study = '2')
```

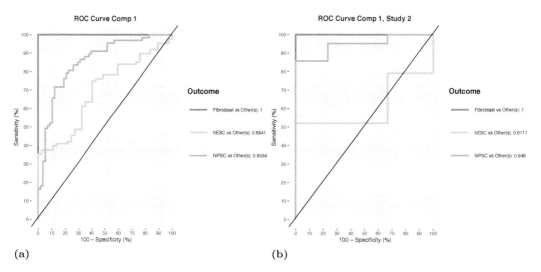

(a)   (b)

FIGURE 14.12: ROC curve and AUC from the MINT sPLS-DA performed on the `stemcells` gene expression data for global and specific studies, averaged across one-vs-all comparisons. Numerical outputs include the AUC and a Wilcoxon test *p*-value for each 'one vs. other' class comparison that are performed per component. This output complements the sPLS-DA performance evaluation but *should not be used for tuning* (as the prediction process in sPLS-DA is based on prediction distances, not a cutoff that maximises specificity and sensitivity as in ROC). The plot suggests that the selected features are more accurate in classifying fibroblasts versus the other cell types, and less accurate in distinguishing hESC versus the other cell types or hiPSC versus the other cell types.

14.5.5.2   Prediction on an external study

We use the `predict()` function to predict the class membership of new test samples from an external study. We provide an example where we set aside a particular study, train the MINT model on the remaining three studies, then predict on the test study. This process exactly reflects the inner workings of the `tune()` and `perf()` functions using LOGOCV.

Here during our model training on the three studies only, we assume we have performed the tuning steps described in this case study to choose `ncomp` and `keepX` (here set to arbitrary values to avoid overfitting):

```
# We predict on study 3
indiv.test <- which(study == "3")
```

```
# We train on the remaining studies, with pre-tuned parameters
mint.splsda.stem2 <- mint.splsda(X = X[-c(indiv.test), ],
                                 Y = Y[-c(indiv.test)],
                                 study = droplevels(study[-c(indiv.test)]),
                                 ncomp = 1,
                                 keepX = 30)
```

```
mint.predict.stem <- predict(mint.splsda.stem2, newdata = X[indiv.test, ],
                             dist = "centroids.dist",
                             study.test = factor(study[indiv.test]))
```

```
# Store class prediction with a model with 1 comp
indiv.prediction <- mint.predict.stem$class$centroids.dist[, 1]
```

```
# The confusion matrix compares the real subtypes with the predicted subtypes
conf.mat <- get.confusion_matrix(truth = Y[indiv.test],
                  predicted = indiv.prediction)
```

```
conf.mat
```

```
##            predicted.as.Fibroblast predicted.as.hESC
## Fibroblast                       3                 0
## hESC                             0                 4
## hiPSC                            2                 2
##            predicted.as.hiPSC
## Fibroblast                  0
## hESC                        4
## hiPSC                       6
```

Here we have considered a trained model with one component, and compared the cell type prediction for the test study 3 with the known cell types. The classification error rate is relatively high, but potentially could be improved with a proper tuning, and a larger number of studies in the training set.

```
# Prediction error rate
(sum(conf.mat) - sum(diag(conf.mat)))/sum(conf.mat)
```

```
## [1] 0.381
```

14.6   Examples of application

The example we provided in this chapter was from Rohart et al. (2017b), but multi-group PLS-DA is not limited to bulk transcriptomics data only, as long as a sufficient number of studies are available.

14.6.1   16S rRNA gene data

In a 16S amplicon microbiome study, the effect of anaerobic digestion inhibition by ammonia and phenol (the outcome) was investigated (Poirier et al., 2020). A MINT sPLS-DA model was trained on two in-house studies, and this model predicted with 90% accuracy ammonia inhibition in two external data sets, as shown in the above case study. The raw counts were filtered then centered log ratio transformed (these steps are described in Sections 4.2 and 12.6).

14.6.2   Single cell transcriptomics

A benchmarking single cell transcriptomics experiment was designed to study the effect of sequencing protocols (Tian et al., 2019). Several data integration methods were compared for single cell transcriptomics data: MNNs (Haghverdi et al., 2018), Scanorama (Hie et al., 2019), and Seurat (Butler et al., 2018) are all unsupervised, whereas scMerge (Lin et al., 2019) has a supervised and unsupervised version. Interestingly, this type of question calls for an integrative method that either corrects or accommodates for batch effects due to protocols, rather than studies from different laboratories. Our further analyses available on our website showed the ability of MINT sPLS-DA to identify relevant gene signatures discriminating cell types.

14.7   To go further

Regression methods such as `mint.pls()` and `mint.spls()` are also available in `mixOmics` albeit more developments are needed for the tuning and performance functions and their application to biological studies. Further examples are provided at `www.mixOmics.org` and details on the MINT method and algorithm are available in Rohart et al. (2017b).

14.8   Summary

MINT is a *P−*integration framework that combines and integrates independent studies generated under similar biological conditions, whilst measuring the same *P* predictors (e.g. genes, taxa). The overarching aim of $P-$integration is to enable data sharing and methods benchmarking by re-using existing data from public databases where independent studies keep accumulating. Multi-group sPLS-(DA) ultimately aims at identifying reproducible biomarker signatures across those studies.

MINT is designed to minimise the systematic, unwanted variation that occurs when combining data from similar studies – i.e. the variation arising from generating data at different locations, times, and from different technological platforms, so as to parse the interesting biological variation. Such analyses have the potential to improve statistical power by increasing the number of samples, and lead to more robust molecular signatures.

MINT is based on multi-group sPLS-DA, a classification framework to discriminate sample groups and identify a set of markers for prediction. A regression framework is also available based on multi-group sPLS, but is still in development. For such a framework, the aim is to integrate predictor and response data matrices and identify either correlated variables from both matrices, or variables from one matrix that best explain the other.

Glossary of terms

- **Individuals, observations or samples**: the experimental units on which information are collected, e.g. patients, cell lines, cells, faecal samples, etc.

- **Variables, predictors**: read-out measured on each sample, e.g. gene (expression), protein or Operational Taxonomic Unit (abundance), weight, etc.

- **Variance**: measures the spread of one variable. In our methods, we estimate the variance of components rather than variable read-outs. A high variance indicates that the data points are spread out from the mean, and from one another (see Section 3.1.2).

- **Covariance**: measures the strength of the relationship between two variables, i.e. whether they covary. A high covariance value indicates a strong relationship, e.g. weight and height in individuals frequently vary roughly in the same way (the heaviest are the tallest). A covariance value has no lower or upper bound (see Section 3.1.3).

- **Correlation**: a standardised version of the covariance that is bounded by -1 and 1 (see Section 3.1.4).

- **Linear combination**: variables are combined by multiplying each of them by a coefficient and adding the results. A linear combination of height and weight could be $2 \times$ weight $-1.5 \times$ height with the coefficients 2 and -1.5 assigned to weight and height, respectively (see Sections 2.4.4 and 5.1.1).

- **Component**: an artificial variable built from a linear combination of the observed variables in a given data set. Variable coefficients are optimally defined based on some statistical criteria. For example in Principal Component Analysis, the coefficients in the (principal) component are defined to maximise the variance of the component (see Section 5.1.1).

- **Loadings**: variable coefficients used to define a component (see Section 5.1.1).

- **Sample plot**: representation of the samples projected in a small space spanned (defined) by the components. Samples coordinates are determined by their component values or scores (see Sections 3.2.3 and 6.1).

- **Correlation circle plot**: representation of the variables in a space spanned by the components. Each variable coordinate is defined as the correlation between the original variable value and each component. A correlation circle plot enables to visualise the correlation between variables - negative or positive correlation, defined by the cosine angle between the centre of the circle and each variable point) and the contribution of each variable to each component – defined by absolute value of the coordinate on each component. For this interpretation, data need to be centred and scaled (by default in most of our methods except PCA, see Section 6.2.2).

- **Unsupervised analysis**: a type of analysis that does not take into account any known sample groups. Our unsupervised methods are exploratory, and include Principal Component Analysis (PCA, Chapter 9), and Canonical Correlation Analysis (CCA, 11).

- **Supervised analysis**: a type of analysis that includes a vector indicating either the class membership of each sample (classification) or a continuous response of each sample (regression). Examples of classification methods covered in this book are PLS Discriminant Analysis (PLS-DA, Chapter 12), DIABLO (Chapter 13), and MINT (Chapter 14). Examples of regression methods are Projection to Latent Structures (PLS, Chapter 10).

Key publications

The methods implemented in `mixOmics` are described in detail in the following publications. A more extensive list can be found at http://mixomics.org/a-propos/publications/.

- **Overview and recent integrative methods**: Rohart et al. (2017a), mixOmics: an R package for 'omics feature selection and multiple data integration, *PLoS Computational Biology* 13(11): e1005752.

- **Graphical outputs for integrative methods** (Correlation circle plots, relevance networks and CIM): González et al. (2012), Insightful graphical outputs to explore relationships between two omics data sets, *BioData Mining* 5:19.

- **sparse PLS**:
  - Lê Cao et al. (2008), A sparse PLS for variable selection when integrating omics data, *Statistical Applications in Genetics and Molecular Biology*, 7:35.
  - Lê Cao et al. (2009), Sparse canonical methods for biological data integration: application to a cross-platform study, *BMC Bioinformatics*, 10:34.

- **sparse PLS-DA**: Lê Cao et al. (2011), Sparse PLS discriminant analysis: biologically relevant feature selection and graphical displays for multiclass problems, *BMC Bioinformatics*, 22:253.

- **Multilevel approach for repeated measurements**: Liquet et al. (2012), A novel approach for biomarker selection and the integration of repeated measures experiments from two assays, *BMC Bioinformatics*, 13:325.

- **sPLS-DA for microbiome data**: Lê Cao et al. (2016), MixMC: Multivariate insights into Microbial communities, *PLoS ONE* 11(8):e0160169.

- **DIABLO**: Singh et al. (2019), DIABLO: an integrative approach for identifying key molecular drivers from multi-omics assays, *Bioinformatics* 35:17.

Bibliography

Abdi, H., Chin, W.W., Esposito Vinzi, V., Russolillo, G., Trinchera, L. (2013). Multi-group pls regression: application to epidemiology. In *New Perspectives in Partial Least Squares and Related Methods*, pages 243–255. Springer.

Aitchison, J. (1982). The statistical analysis of compositional data. *Journal of the Royal Statistical Society. Series B (Methodological)*, 44(2):pages 139–177.

Ambroise, C. and McLachlan, G. J. (2002). Selection bias in gene extraction in tumour classification on basis of microarray gene expression data. *Proc. Natl. Acad. Sci. USA*, 99(1):6562–6566.

Argelaguet, R., Velten, B., Arnol, D., Dietrich, S., Zenz, T., Marioni, J. C., Buettner, F., Huber, W., and Stegle, O. (2018). Multi-omics factor analysis–a framework for unsupervised integration of multi-omics data sets. *Molecular Systems Biology*, 14(6):e8124.

Arumugam, M., Raes, J., Pelletier, E., Le Paslier, D., Yamada, T., Mende, D. R., Fernandes, G. R., Tap, J., Bruls, T., Batto, J.-M., et al. (2011). Enterotypes of the human gut microbiome. *Nature*, 473(7346):174.

Baumer, B. and Udwin, D. (2015). R markdown. *Wiley Interdisciplinary Reviews: Computational Statistics*, 7(3):167–177.

Blasco-Moreno, A., Pérez-Casany, M., Puig, P., Morante, M., and Castells, E. (2019). What does a zero mean? Understanding false, random and structural zeros in ecology. *Methods in Ecology and Evolution*, 10(7):949–959.

Bodein, A., Chapleur, O., Cao, K.-A. L., and Droit, A. (2020). *timeOmics: Time-Course Multi-Omics Data Integration*. R package version 1.0.0.

Bodein, A., Chapleur, O., Droit, A., and Lê Cao, K.-A. (2019). A generic multivariate framework for the integration of microbiome longitudinal studies with other data types. *Frontiers in Genetics*, 10.

Boulesteix, A.-L. (2004). Pls dimension reduction for classification with microarray data. *Statistical Applications in Genetics And Molecular Biology*, 3(1):1–30.

Boulesteix, A. and Strimmer, K. (2005). Predicting transcription factor activities from combined analysis of microarray and chip data: a partial least squares approach. *Theoretical Biology and Medical Modelling*, 2(23).

Boulesteix, A. and Strimmer, K. (2007). Partial least squares: a versatile tool for the analysis of high-dimensional genomic data. *Briefings in Bioinformatics*, 8(1):32.

Bushel, P. R., Wolfinger, R. D., and Gibson, G. (2007). Simultaneous clustering of gene expression data with clinical chemistry and pathological evaluations reveals phenotypic prototypes. *BMC Systems Biology*, 1(1):15.

Butler, A., Hoffman, P., Smibert, P., Papalexi, E., and Satija, R. (2018). Integrating single-cell transcriptomic data across different conditions, technologies, and species. *Nature Biotechnology*, 36(5):411–420.

Butte, A. J., Tamayo, P., Slonim, D., Golub, T. R., and Kohane, I. S. (2000). Discovering functional relationships between RNA expression and chemotherapeutic susceptibility using relevance networks. *Proceedings of the National Academy of Sciences of the USA*, 97:12182–12186.

Bylesjö, M., Rantalainen, M., Cloarec, O., Nicholson, J. K., Holmes, E., and Trygg, J. (2006). OPLS discriminant analysis: combining the strengths of PLS-DA and SIMCA classification. *Journal of Chemometrics: A Journal of the Chemometrics Society*, 20(8–10):341–351.

Calude, C. S. and Longo, G. (2017). The deluge of spurious correlations in big data. *Foundations of Science*, 22(3):595–612.

Cancer Genome Atlas Network et al. (2012). Comprehensive molecular portraits of human breast tumours. *Nature*, 490(7418):61–70.

Caporaso, J. G., Kuczynski, J., Stombaugh, J., Bittinger, K., Bushman, F. D., Costello, E. K., Fierer, N., Pena, A. G., Goodrich, J. K., Gordon, J. I., et al. (2010). QIIME allows analysis of high-throughput community sequencing data. *Nature Methods*, 7(5):335.

Chin, M. H., Mason, M. J., Xie, W., Volinia, S., Singer, M., Peterson, C., Ambartsumyan, G., Aimiuwu, O., Richter, L., Zhang, J., et al. (2009). Induced pluripotent stem cells and embryonic stem cells are distinguished by gene expression signatures. *Cell Stem Cell*, 5(1):111–123.

Chun, H. and Keleş, S. (2010). Sparse partial least squares regression for simultaneous dimension reduction and variable selection. *Journal of the Royal Statistical Society: Series B (Statistical Methodology)*, 72(1):3–25.

Chung, D., Chun, H., and S., K. (2020). *spls: Sparse Partial Least Squares (SPLS) Regression and Classification*. R package version 2:2–3.

Chung, D. and Keles, S. (2010). Sparse Partial Least Squares Classification for high dimensional data. *Statistical Applications in Genetics and Molecular Biology*, 9(1):17.

Combes, S., González, I., Déjean, S., Baccini, A., Jehl, N., Juin, H., Cauquil, L., and BÈatrice Gabinaud, François Lebas, C. L. (2008). Relationships between sensorial and physicochemical measurements in meat of rabbit from three different breeding systems using canonical correlation analysis. *Meat Science*, Meat Science, 80(3):835–841.

Comon, P. (1994). Independent component analysis, a new concept? *Signal Processing*, 36(3):287–314.

Csala, A. (2017). *sRDA: Sparse Redundancy Analysis*. R package version 1.0.0.

Csala, A., Voorbraak, F. P., Zwinderman, A. H., and Hof, M. H. (2017). Sparse redundancy analysis of high-dimensional genetic and genomic data. *Bioinformatics*, 33(20):3228–3234.

Csardi, G., Nepusz, T., et al. (2006). The igraph software package for complex network research. *InterJournal, Complex Systems*, 1695(5):1–9.

De La Fuente, A., Bing, N., Hoeschele, I., and Mendes, P. (2004). Discovery of meaningful associations in genomic data using partial correlation coefficients. *Bioinformatics*, 20(18):3565–3574.

De Tayrac, M., Lê, S., Aubry, M., Mosser, J., and Husson, F. (2009). Simultaneous analysis of distinct omics data sets with integration of biological knowledge: Multiple factor analysis approach. *BMC Genomics*, 10(1):1–17.

De Vries, A. and Meys, J. (2015). *R for Dummies*. John Wiley & Sons.

Dray, S., Dufour, A.-B., et al. (2007). The ade4 package: implementing the duality diagram for ecologists. *Journal of Statistical Software*, 22(4):1–20.

Drier, Y., Sheffer, M., and Domany, E. (2013). Pathway-based personalized analysis of cancer. *Proceedings of the National Academy of Sciences*, 110(16):6388–6393.

Efron, B. and Tibshirani, R. (1997). Improvements on cross-validation: the 632+ bootstrap method. *Journal of the American Statistical Association*, 92(438):548–560.

Eisen, M. B., Spellman, P. T., Brown, P. O., and Botstein, D. (1998). Cluster analysis and display of genome-wide expression patterns. *Proceeding of the National Academy of Sciences of the USA*, 95:14863–14868.

Escudié, F., Auer, L., Bernard, M., Mariadassou, M., Cauquil, L., Vidal, K., Maman, S., Hernandez-Raquet, G., Combes, S., and Pascal, G. (2017). Frogs: find, rapidly, OTUs with galaxy solution. *Bioinformatics*, 34(8):1287–1294.

Eslami, A., Qannari, E. M., Kohler, A., and Bougeard, S. (2014). Algorithms for multi-group pls. *Journal of Chemometrics*, 28(3):192–201.

Féraud, B., Munaut, C., Martin, M., Verleysen, M., and Govaerts, B. (2017). Combining strong sparsity and competitive predictive power with the L-sOPLS approach for biomarker discovery in metabolomics. *Metabolomics*, 13(11):130.

Fernandes, A. D., Reid, J. N., Macklaim, J. M., McMurrough, T. A., Edgell, D. R., and Gloor, G. B. (2014). Unifying the analysis of high-throughput sequencing datasets: characterizing RNA-seq, 16s rrna gene sequencing and selective growth experiments by compositional data analysis. *Microbiome*, 2(1):1.

Fornell, C., Barclay, D. W., and Rhee, B.-D. (1988). A Model and Simple Iterative Algorithm For Redundancy Analysis. *Multivariate Behavioral Research*, 23(3):349–360.

Friedman, J. H. (1989). Regularized discriminant analysis. *Journal of the American Statistical Association*, 84:165–175.

Friedman, J., Hastie, T., Höfling, H., Tibshirani, R., et al. (2007). Pathwise coordinate optimization. *The Annals of Applied Statistics*, 1(2):302–332.

Friedman, J., Hastie, T., and Tibshirani, R. (2001). *The Elements of Statistical Learning*, volume 1. Springer, Series in statistics, New York.

Friedman, J., Hastie, T., and Tibshirani, R. (2010). Regularization paths for generalized linear models via coordinate descent. *Journal of Statistical Software*, 33(1):1.

Gagnon-Bartsch, J. A. and Speed, T. P. (2012). Using control genes to correct for unwanted variation in microarray data. *Biostatistics*, 13(3):539–552.

Gaude, E., Chignola, F., Spiliotopoulos, D., Spitaleri, A., Ghitti, M., Garcìa-Manteiga, J. M., Mari, S., and Musco, G. (2013). muma, an R package for metabolomics univariate and multivariate statistical analysis. *Current Metabolomics*, 1(2):180–189.

Gittins, R. (1985). *Canonical Analysis: A Review with Applications in Ecology*. Springer-Verlag.

Gligorijević, V. and Pržulj, N. (2015). Methods for biological data integration: perspectives and challenges. *Journal of the Royal Society Interface*, 12(112):20150571.

Gloor, G. B., Macklaim, J. M., Pawlowsky-Glahn, V., and Egozcue, J. J. (2017). Microbiome datasets are compositional: and this is not optional. *Frontiers in Microbiology*, 8:2224.

Wilkinson J. H., Reinsch C. (1971). Singular value decomposition and least squares solutions. In *Linear Algebra*, pages 134–151. Springer.

Golub, G. and Van Loan, C. (1996). *Matrix Computations*. Johns Hopkins University Press.

González, I., Déjean, S., Martin, P. G., and Baccini, A. (2008). CCA: An R package to extend canonical correlation analysis. *Journal of Statistical Software*, 23(12):1–14.

González, I., Déjean, S., Martin, P., Gonçalves, O., Besse, P., and Baccini, A. (2009). Highlighting relationships between heterogeneous biological data through graphical displays based on regularized canonical correlation analysis. *Journal of Biological Systems*, 17(02):173–199.

González, I., Lê Cao, K.-A., Davis, M. J., Déjean, S., et al. (2012). Visualising associations between paired 'omics' data sets. *BioData Mining*, 5(1):19.

Günther, O. P., Chen, V., Freue, G. C., Balshaw, R. F., Tebbutt, S. J., Hollander, Z., Takhar, M., McMaster, W. R., McManus, B. M., Keown, P. A., et al. (2012). A computational pipeline for the development of multi-marker bio-signature panels and ensemble classifiers. *BMC bioinformatics*, 13(1):326.

Haghverdi, L., Lun, A. T., Morgan, M. D., and Marioni, J. C. (2018). Batch effects in single-cell RNA-sequencing data are corrected by matching mutual nearest neighbors. *Nature Biotechnology*, 36(5):421–427.

Hardoon, D. R. and Shawe-Taylor, J. (2011). Sparse canonical correlation analysis. *Machine Learning*, 83(3):331–353.

Hastie, T. and Stuetzle, W. (1989). Principal curves. *Journal of the American Statistical Association*, 84(406):502–516.

Hawkins, D. M. (2004). The problem of overfitting. *Journal of Chemical Information and Computer Sciences*, 44(1):1–12.

Hervé, Maxime. "RVAideMemoire: testing and plotting procedures for biostatistics. R package version 0.9-69 3 (2018).

Hervé, M. R., Nicolè, F., and Lê Cao, K.-A. (2018). Multivariate analysis of multiple datasets: a practical guide for chemical ecology. *Journal of Chemical Ecology*, 44(3):215–234.

Hie, B., Bryson, B., and Berger, B. (2019). Efficient integration of heterogeneous single-cell transcriptomes using Scanorama. *Nature Biotechnology*, 37(6):685–691.

Hoerl, A. E. (1964). Ridge analysis. In Chemical engineering progress symposium series (Vol. 60, No. 67–77, p. 329).

Hoerl, A. E. and Kennard, R. W. (1970). Ridge regression: biased estimation for non-orthogonal problems. *Technometrics*, 12:55–67.

Hornung, R., Boulesteix, A.-L., and Causeur, D. (2016). Combining location-and-scale batch effect adjustment with data cleaning by latent factor adjustment. *BMC Bioinformatics*, 17(1):27.

Hoskuldsson, A. (1988). PLS regression methods. *Journal of Chemometrics*, 2(3):211–228.

Hotelling, H. (1936). Relations between two sets of variates. *Biometrika*, 28:321–377.

Huang, S., Chaudhary, K., and Garmire, L. X. (2017). More is better: recent progress in multi-omics data integration methods. *Frontiers in Genetics*, 8:84.

Hyvärinen, A. and Oja, E. (2001). Independent component analysis: algorithms and applications. *Neural Networks*, 13(4):411–430.

Indahl, U. G., Liland, K. H., and Næs, T. (2009). Canonical partial least squares – a unified pls approach to classification and regression problems. *Journal of Chemometrics: A Journal of the Chemometrics Society*, 23(9):495–504.

Indahl, U. G., Martens, H., and Næs, T. (2007). From dummy regression to prior probabilities in pls-da. *Journal of Chemometrics: A Journal of the Chemometrics Society*, 21(12):529–536.

James, G., Witten, D., Hastie, T., and Tibshirani, R. (2013). *An Introduction to Statistical Learning*, volume 112. Springer.

Johnson, W. E., Li, C., and Rabinovic, A. (2007). Adjusting batch effects in microarray expression data using empirical Bayes methods. *Biostatistics*, 8(1):118–127.

Jolliffe, I. (2005). *Principal component analysis*. Wiley Online Library.

Karlis, D., Saporta, G., and Spinakis, A. (2003). A simple rule for the selection of principal components. *Communications in Statistics-Theory and Methods*, 32(3):643–666.

Khan, J., Wei, J. S., Ringner, M., Saal, L. H., Ladanyi, M., Westermann, F., Berthold, F., Schwab, M., Antonescu, C. R., Peterson, C., et al. (2001). Classification and diagnostic prediction of cancers using gene expression profiling and artificial neural networks. *Nature Medicine*, 7(6):673–679.

Kim, Y., Kwon, S., and Choi, H. (2012). Consistent model selection criteria on high dimensions. *Journal of Machine Learning Research*, 13(Apr):1037–1057.

Kunin, V., Engelbrektson, A., Ochman, H., and Hugenholtz, P. (2010). Wrinkles in the rare biosphere: pyrosequencing errors can lead to artificial inflation of diversity estimates. *Environmental Microbiology*, 12(1):118–123.

Lazzeroni, L. and Ray, A. (2012). The cost of large numbers of hypothesis tests on power, effect size and sample size. *Molecular Psychiatry*, 17(1):108.

Lê Cao, K.-A., Boitard, S., and Besse, P. (2011). Sparse PLS Discriminant Analysis: biologically relevant feature selection and graphical displays for multiclass problems. *BMC Bioinformatics*, 12(1):253.

Lê Cao, K.-A., Costello, M.-E., Chua, X.-Y., Brazeilles, R., and Rondeau, P. (2016). MixMC: Multivariate insights into microbial communities. *PloS One*, 11(8):e0160169.

Lê Cao, K.-A., Martin, P. G., Robert-Granié, C., and Besse, P. (2009). Sparse canonical methods for biological data integration: application to a cross-platform study. *BMC Bioinformatics*, 10(1):34.

Lê Cao, K.-A., Rohart, F., McHugh, L., Korn, O., and Wells, C. A. (2014). YuGene: a simple approach to scale gene expression data derived from different platforms for integrated analyses. *Genomics*, 103(4):239–251.

Lê Cao, K., Rossouw, D., Robert-Granié, C., Besse, P., et al. (2008). A sparse PLS for variable selection when integrating omics data. *Statistical Applications in Genetics and Molecular Biology*, 7(35).

Lee, A. H., Shannon, C. P., Amenyogbe, N., Bennike, T. B., Diray-Arce, J., Idoko, O. T., Gill, E. E., Ben-Othman, R., Pomat, W. S., Van Haren, S. D., et al. (2019). Dynamic molecular changes during the first week of human life follow a robust developmental trajectory. *Nature Communications*, 10(1):1092.

Leek, J. T. and Storey, J. D. (2007). Capturing heterogeneity in gene expression studies by surrogate variable analysis. *PLoS Genetics*, 3(9):e161.

Lenth, R. V. (2001). Some practical guidelines for effective sample size determination. *The American Statistician*, 55(3):187–193.

Leurgans, S. E., Moyeed, R. A., and Silverman, B. W. (1993). Canonical correlation analysis when the data are curves. *Journal of the Royal Statistical Society. Series B*, 55:725–740.

Li, Z., Safo, S. E., and Long, Q. (2017). Incorporating biological information in sparse principal component analysis with application to genomic data. *BMC Bioinformatics*, 18(1):332.

Lin, Y., Ghazanfar, S., Wang, K. Y., Gagnon-Bartsch, J. A., Lo, K. K., Su, X., Han, Z.-G., Ormerod, J. T., Speed, T. P., Yang, P., and JYH, Y. (2019). scMerge leverages factor analysis, stable expression, and pseudoreplication to merge multiple single-cell RNA-seq datasets. *Proceedings of the National Academy of Sciences of the United States of America*, 116(20):9775.

Liquet, B., de Micheaux, P. L., Hejblum, B. P., and Thiébaut, R. (2016). Group and sparse group partial least square approaches applied in genomics context. *Bioinformatics*, 32(1):35–42.

Liquet, B., Lafaye de Micheaux, P., and Broc, C. (2017). *sgPLS: Sparse Group Partial Least Square Methods*. R package version 1.7.

Liquet, B., Lê Cao, K.-A., Hocini, H., and Thiébaut, R. (2012). A novel approach for biomarker selection and the integration of repeated measures experiments from two assays. *BMC Bioinformatics*, 13:325.

Listgarten, J., Kadie, C., Schadt, E. E., and Heckerman, D. (2010). Correction for hidden confounders in the genetic analysis of gene expression. *Proceedings of the National Academy of Sciences*, 107(38):16465–16470.

Liu, C., Srihari, S., Lal, S., Gautier, B., Simpson, P. T., Khanna, K. K., Ragan, M. A., and Lê Cao, K.-A. (2016). Personalised pathway analysis reveals association between DNA repair pathway dysregulation and chromosomal instability in sporadic breast cancer. *Molecular Oncology*, 10(1):179–193.

Lock, E. F., Hoadley, K. A., Marron, J. S., and Nobel, A. B. (2013). Joint and individual variation explained (jive) for integrated analysis of multiple data types. *The Annals of Applied Statistics*, 7(1):523.

Lohmöller, J.-B. (2013). *Latent Variable Path Modeling with Partial Least Squares*. Springer Science & Business Media.

Lonsdale, J., Thomas, J., Salvatore, M., Phillips, R., Lo, E., Shad, S., Hasz, R., Walters, G., Garcia, F., Young, N., et al. (2013). The genotype-tissue expression (GTEx) project. *Nature Genetics*, 45(6):580.

Lorber, A., Wangen, L., and Kowalski, B. (1987). A theoretical foundation for the PLS algorithm. *Journal of Chemometrics*, 1(19-31):13.

Lovell, D., Pawlowsky-Glahn, V., Egozcue, J. J., Marguerat, S., and Bähler, J. (2015). Proportionality: a valid alternative to correlation for relative data. *PLoS Computational Biology*, 11(3):e1004075.

Luecken, M. D. and Theis, F. J. (2019). Current best practices in single-cell RNA-seq analysis: a tutorial. *Molecular Systems Biology*, 15(6).

Ma, X., Chen, T., and Sun, F. (2013). Integrative approaches for predicting protein function and prioritizing genes for complex phenotypes using protein interaction networks. *Briefings in Bioinformatics*, 15(5):685–698.

MacKay, R. J. and Oldford, R. W. (2000). Scientific method, statistical method and the speed of light. *Statistical Science*, pages 254–278.

Mardia, K. V., Kent, J. T., and Bibby, J. M. (1979). *Multivariate Analysis*. Academic Press.

Martin, P., Guillou, H., Lasserre, F., Déjean, S., Lan, A., Pascussi, J.-M., San Cristobal, M., Legrand, P., Besse, P., and Pineau, T. (2007). Novel aspects of PPARalpha-mediated regulation of lipid and xenobiotic metabolism revealed through a nutrigenomic study. *Hepatology*, 54:767–777.

Meinshausen, N. and Bühlmann, P. (2010). Stability selection. *Journal of the Royal Statistical Society: Series B (Statistical Methodology)*, 72(4):417–473.

Mizuno, H., Ueda, K., Kobayashi, Y., Tsuyama, N., Todoroki, K., Min, J. Z., and Toyo'oka, T. (2017). The great importance of normalization of LC–MS data for highly-accurate non-targeted metabolomics. *Biomedical Chromatography*, 31(1):e3864.

Muñoz-Romero, S., Arenas-García, J., and Gómez-Verdejo, V. (2015). Sparse and kernel OPLS feature extraction based on eigenvalue problem solving. *Pattern Recognition*, 48(5):1797–1811.

Murdoch, D. and Chow, E. (1996). A graphical display of large correlation matrices. *The American Statistician*, 50(2):178–180.

Newman, A. M. and Cooper, J. B. (2010). Lab-specific gene expression signatures in pluripotent stem cells. *Cell Stem Cell*, 7(2):258–262.

Nguyen, D. V. and Rocke, D. M. (2002a). Multi-class cancer classification via partial least squares with gene expression profiles. *Bioinformatics*, 18(9):1216–1226.

Nguyen, D. V. and Rocke, D. M. (2002b). Tumor classification by partial least squares using microarray gene expression data. *Bioinformatics*, 18(1):39.

Nichols, J. D., Oli, M. K., Kendall, W. L., and Boomer, G. S. (2021). Opinion: A better approach for dealing with reproducibility and replicability in science. *Proceedings of the National Academy of Sciences*, 118(7).

O'Connell, M. J. and Lock, E. F. (2016). R. jive for exploration of multi-source molecular data. *Bioinformatics*, 32(18):2877–2879.

Pan, W., Xie, B., and Shen, X. (2010). Incorporating predictor network in penalized regression with application to microarray data. *Biometrics*, 66(2):474–484.

Parkhomenko, E., Tritchler, D., and Beyene, J. (2009). Sparse canonical correlation analysis with application to genomic data integration. *Statistical Applications in Genetics and Molecular Biology*, 8(1):1–34.

Poirier, S., Déjean, S., Midoux, C., Lê Cao, K.-A., and Chapleur, O. (2020). Integrating independent microbial studies to build predictive models of anaerobic digestion inhibition. *Bioresource Technology*, 316(123952).

Prasasya, R. D., Vang, K. Z., and Kreeger, P. K. (2012). A multivariate model of ErbB network composition predicts ovarian cancer cell response to canertinib. *Biotechnology and Bioengineering*, 109(1):213–224.

Ricard Argelaguet, Britta Velten, Damien Arnol, Florian Buettner, Wolfgang Huber and Oliver Stegle (2020). *MOFA: Multi-Omics Factor Analysis (MOFA)*. R package version 1.4.0.

Richardson, S., Tseng, G., and Sun, W. (2016). Statistical methods in integrative genomics. *Annual Review of Statistics and Its Application*, 3(1):181–209.

Ritchie, M. D., de Andrade, M., and Kuivaniemi, H. (2015). The foundation of precision medicine: integration of electronic health records with genomics through basic, clinical, and translational research. *Frontiers in genetics*, 6:104.

Robinson, M. D. and Oshlack, A. (2010). A scaling normalization method for differential expression analysis of RNA-seq data. *Genome Biology*, 11(3):R25.

Roemer, E. (2016). A tutorial on the use of PLS path modeling in longitudinal studies. *Industrial Management & Data Systems*, Vol. 116 No. 9, pp. 1901–1921.

Rohart, F., Gautier, B., Singh, A., and Lê Cao, K.-A. (2017a). mixOmics: An r package for 'omics feature selection and multiple data integration. *PLoS Computational Biology*, 13(11):e1005752.

Rohart, F., Matigian, N., Eslami, A., Stephanie, B., and Lê Cao, K.-A. (2017b). Mint: a multivariate integrative method to identify reproducible molecular signatures across independent experiments and platforms. *BMC Bioinformatics*, 18(1):128.

Ruiz-Perez, D., Guan, H., Madhivanan, P., Mathee, K., and Narasimhan, G. (2020). So you think you can pls-da? *BMC Bioinformatics*, 21(1):1–10.

Saccenti, E., Hoefsloot, H. C., Smilde, A. K., Westerhuis, J. A., and Hendriks, M. M. (2014). Reflections on univariate and multivariate analysis of metabolomics data. *Metabolomics*, 10(3):0.

Saccenti, E. and Timmerman, M. E. (2016). Approaches to sample size determination for multivariate data: Applications to pca and pls-da of omics data. *Journal of Proteome Research*, 15(8):2379–2393.

Saelens, W., Cannoodt, R., and Saeys, Y. (2018). A comprehensive evaluation of module detection methods for gene expression data. *Nature Communications*, 9(1):1–12.

Saporta, G. (2006). *Probabilités analyse des données et statistique*. Technip.

Schafer, J., Opgen-Rhein, R., Zuber, V., Ahdesmaki, M., Silva, A. P. D., and Strimmer., K. (2017). *corpcor: Efficient Estimation of Covariance and (Partial) Correlation*. R package version 1.6.9.

Schäfer, J. and Strimmer, K. (2005). A shrinkage approach to large-scale covariance matrix estimation and implications for functional genomics. *Statistical Applications in Genetics and Molecular Biology*, 4(1).

Scherf, U., Ross, D. T., Waltham, M., Smith, L. H., Lee, J. K., Tanabe, L., Kohn, K. W., Reinhold, W. C., Myers, T. G., Andrews, D. T., et al. (2000). A gene expression database for the molecular pharmacology of cancer. *Nature Genetics*, 24(3):236–244.

Shen, H. and Huang, J. Z. (2008). Sparse principal component analysis via regularized low rank matrix approximation. *Journal of Multivariate Analysis*, 99(6):1015–1034.

Shmueli, G. (2010). To explain or to predict? *Statistical Science*, 25(3):289–310.

Sill, M., Saadati, M., and Benner, A. (2015). Applying stability selection to consistently estimate sparse principal components in high-dimensional molecular data. *Bioinformatics*, 31(16):2683–2690.

Simon, N., Friedman, J., Hastie, T., and Tibshirani, R. (2013). A sparse-group lasso. *Journal of Computational and Graphical Statistics*, 22(2):231–245.

Simon, N. and Tibshirani, R. (2012). Standardization and the group lasso penalty. *Statistica Sinica*, 22(3):983.

Sims, A. H., Smethurst, G. J., Hey, Y., Okoniewski, M. J., Pepper, S. D., Howell, A., Miller, C. J., and Clarke, R. B. (2008). The removal of multiplicative, systematic bias allows integration of breast cancer gene expression datasets – improving meta-analysis and prediction of prognosis. *BMC medical genomics*, 1(1):42.

Singh, A., Shannon, C. P., Gautier, B., Rohart, F., Vacher, M., Tebbutt, S. J., and Lê Cao, K.-A. (2019). Diablo: an integrative approach for identifying key molecular drivers from multi-omics assays. *Bioinformatics*, 35(17):3055–3062.

Smit, S., van Breemen, M. J., Hoefsloot, H. C., Smilde, A. K., Aerts, J. M., and De Koster, C. G. (2007). Assessing the statistical validity of proteomics based biomarkers. *Analytica Chimica Acta*, 592(2):210–217.

Smyth, G. K. (2005). Limma: linear models for microarray data. In *Bioinformatics and Computational Biology Solutions Using R and Bioconductor*, pages 397–420. Springer.

Sompairac, N., Nazarov, P. V., Czerwinska, U., Cantini, L., Biton, A., Molkenov, A., Zhumadilov, Z., Barillot, E., Radvanyi, F., Gorban, A., et al. (2019). Independent component analysis for unraveling the complexity of cancer omics datasets. *International Journal of Molecular Sciences*, 20(18):4414.

Sørlie, T., Perou, C. M., Tibshirani, R., Aas, T., Geisler, S., Johnsen, H., Hastie, T., Eisen, M. B., Van De Rijn, M., Jeffrey, S. S., et al. (2001). Gene expression patterns of breast carcinomas distinguish tumor subclasses with clinical implications. *Proceedings of the National Academy of Sciences*, 98(19):10869–10874.

Ståhle, L. and Wold, S. (1987). Partial least squares analysis with cross-validation for the two-class problem: a Monte Carlo study. *Journal of Chemometrics*, 1(3):185–196.

Stein-O'Brien, G. L., Arora, R., Culhane, A. C., Favorov, A. V., Garmire, L. X., Greene, C. S., ... and Fertig, E. J. (2018). Enter the matrix: factorization uncovers knowledge from omics. *Trends in Genetics*, 34(10):790–805.

Straube, J., Gorse, A. D., PROOF Centre of Excellence Team, Huang, B. E., and Lê Cao, K. A. (2015). A linear mixed model spline framework for analysing time course 'omics' data. *PLoS ONE*, 10(8):e0134540.

Susin, A., Wang, Y., Lê Cao, K.-A., and Calle, M. L. (2020). Variable selection in microbiome compositional data analysis. *NAR Genomics and Bioinformatics*, 2(2):lqaa029.

Szakács, G., Annereau, J.-P., Lababidi, S., Shankavaram, U., Arciello, A., Bussey, K. J., Reinhold, W., Guo, Y., Kruh, G. D., Reimers, M., et al. (2004). Predicting drug sensitivity and resistance: profiling ABC transporter genes in cancer cells. *Cancer Cell*, 6(2):129–137.

Tan, Y., Shi, L., Tong, W., Gene Hwang, G., and Wang, C. (2004). Multi-class tumor classification by discriminant partial least squares using microarray gene expression data and assessment of classification models. *Computational Biology and Chemistry*, 28(3):235–243.

Tapp, H. S. and Kemsley, E. K. (2009). Notes on the practical utility of OPLS. *TrAC Trends in Analytical Chemistry*, 28(11):1322–1327.

Tenenhaus, M. (1998). *La régression PLS: théorie et pratique*. Editions Technip.

Tenenhaus, A. and Guillemot, V. (2017). RGCCA: Regularized and sparse generalized canonical correlation analysis for multiblock data. *R package version*, 2(2).

Tenenhaus, A., Philippe, C., and Frouin, V. (2015). Kernel generalized canonical correlation analysis. *Computational Statistics & Data Analysis*, 90:114–131.

Tenenhaus, A., Philippe, C., Guillemot, V., Le Cao, K. A., Grill, J., and Frouin, V. (2014). Variable selection for generalized canonical correlation analysis. *Biostatistics*, 15(3):569–583.

Tenenhaus, A. and Tenenhaus, M. (2011). Regularized generalized canonical correlation analysis. *Psychometrika*, 76(2):257–284.

Tenenhaus, M., Tenenhaus, A., and Groenen, P. J. (2017). Regularized generalized canonical correlation analysis: a framework for sequential multiblock component methods. *Psychometrika*, 82(3):737–777.

Thévenot, E. A., Roux, A., Xu, Y., Ezan, E., and Junot, C. (2015). Analysis of the human adult urinary metabolome variations with age, body mass index, and gender by implementing a comprehensive workflow for univariate and opls statistical analyses. *Journal of Proteome Research*, 14(8):3322–3335.

Tian, L., Dong, X., Freytag, S., Lê Cao, K.-A., Su, S., JalalAbadi, A., Amann-Zalcenstein, D., Weber, T. S., Seidi, A., Jabbari, J. S., et al. (2019). Benchmarking single cell RNA-sequencing analysis pipelines using mixture control experiments. *Nature Methods*, 16(6):479–487.

Tibshirani, R. (1996). Regression shrinkage and selection via the lasso. *Journal of the Royal Statistical Society: Series B (Methodological)*, 58(1):267–288.

Trygg, J. and Lundstedt, T. (2007). *Chemometrics Techniques for Metabonomics*, pages 171–200. Elsevier.

Trygg, J. and Wold, S. (2002). Orthogonal projections to latent structures (O-PLS). *Journal of Chemometrics*, 16(3):119–128.

Tseng, G. C., Ghosh, D., and Feingold, E. (2012). Comprehensive literature review and statistical considerations for microarray meta-analysis. *Nucleic Acids Research*, 40(9):3785–3799.

Umetri, A. (1996). SIMCA-P for windows, Graphical Software for Multivariate Process Modeling. Umea, Sweden.

Välikangas, T., Suomi, T., and Elo, L. L. (2016). A systematic evaluation of normalization methods in quantitative label-free proteomics. *Briefings in Bioinformatics*, 19(1):1–11.

van den Wollenberg, A. L. (1977). Redundancy analysis an alternative for canonical correlation analysis. *Psychometrika*, 42(2):207–219.

van Gerven, M. A., Chao, Z. C., and Heskes, T. (2012). On the decoding of intracranial data using sparse orthonormalized partial least squares. *Journal of Neural Engineering*, 9(2):026017.

Vinod, H. D. (1976). Canonical ridge and econometrics of joint production. *Journal of Econometrics*, 6:129–137.

Waaijenborg, S., de Witt Hamer, P. C. V., and Zwinderman, A. H. (2008). Quantifying the association between gene expressions and DNA-markers by penalized canonical correlation analysis. *Statistical Applications in Genetics and Molecular Biology*, 7(1).

Wang, T., Guan, W., Lin, J., Boutaoui, N., Canino, G., Luo, J., Celedón, J. C., and Chen, W. (2015). A systematic study of normalization methods for Infinium 450k methylation data using whole-genome bisulfite sequencing data. *Epigenetics*, 10(7):662–669.

Wang, Y. and Lê Cao, K.-A. (2019). Managing batch effects in microbiome data. *Briefings in Bioinformatics*, 21:1477–4054.

Wang, Y. and Lê Cao, K.-A. (2020). A multivariate method to correct for batch effects in microbiome data. *bioRxiv*.

Wang, Z. and Xu, X. (2021). Testing high dimensional covariance matrices via posterior bayes factor. *Journal of Multivariate Analysis*, 181:104674.

Wegelin, J. A. et al. (2000). A survey of partial least squares (pls) methods, with emphasis on the two-block case. Technical report, University of Washington, Tech. Rep.

Weinstein, J. N., Kohn, K. W., Grever, M. R., Viswanadhan, V. N., Rubinstein, L. V., Monks, A. P., Scudiero, D. A., Welch, L., Koutsoukos, A. D., Chiausa, A. J., et al. (1992). Neural computing in cancer drug development: predicting mechanism of action. *Science*, 258(5081):447–451.

Weinstein, J., Myers, T., Buolamwini, J., Raghavan, K., Van Osdol, W., Licht, J., Viswanadhan, V., Kohn, K., Rubinstein, L., Koutsoukos, A., et al. (1994). Predictive statistics and artificial intelligence in the us national cancer institute's drug discovery program for cancer and aids. *Stem Cells*, 12(1):13–22.

Weinstein, J. N., Myers, T. G., O'Connor, P. M., Friend, S. H., Fornace Jr, A. J., Kohn, K. W., Fojo, T., Bates, S. E., Rubinstein, L. V., Anderson, N. L., Buolamwini, J. K., van Osdol, W. W., Monks, A. P., Scudiero, D. A., Sausville, E. A., Zaharevitz, D. W., Bunow, B., Viswanadhan, V. N., Johnson, G. S., Wittes, R. E., and Paull, K. D. (1997). An information-intensive approach to the molecular pharmacology of cancer. *Science*, 275:343–349.

Wells, C. A., Mosbergen, R., Korn, O., Choi, J., Seidenman, N., Matigian, N. A., Vitale, A. M., and Shepherd, J. (2013). Stemformatics: visualisation and sharing of stem cell gene expression. *Stem Cell Research*, 10(3):387–395.

Westerhuis, J. A., Hoefsloot, H. C., Smit, S., Vis, D. J., Smilde, A. K., van Velzen, E. J., van Duijnhoven, J. P., and van Dorsten, F. A. (2008). Assessment of PLSDA cross validation. *Metabolomics*, 4(1):81–89.

Westerhuis, J. A., van Velzen, E. J., Hoefsloot, H. C., and Smilde, A. K. (2010). Multivariate paired data analysis: multilevel PLSDA versus OPLSDA. *Metabolomics*, 6(1):119–128.

Wilms, I. and Croux, C. (2015). Sparse canonical correlation analysis from a predictive point of view. *Biometrical Journal*, 57(5):834–851.

Wilms, I. and Croux, C. (2016). Robust sparse canonical correlation analysis. *BMC Systems Biology*, 10(1):72.

Witten, D. and Tibshirani, R. (2020). *PMA: Penalized Multivariate Analysis*. R package version 1.2.1.

Witten, D., Tibshirani, R., Gross, S., and Narasimhan, B. (2013). PMA: Penalized multivariate analysis. R Package Version, 1(9).

Witten, D. M., Tibshirani, R., and Hastie, T. (2009). A penalized matrix decomposition, with applications to sparse principal components and canonical correlation analysis. *Biostatistics*, 10(3):515–534.

Wold, H. (1966). *Estimation of Principal Components and Related Models by Iterative Least Squares*, pages 391–420. New York: Academic Press.

Wold, H. (1982). Soft modeling: the basic design and some extensions. Systems under indirect observation, 2:343.

Wold, S., Sjöström, M., and Eriksson, L. (2001). Pls-regression: a basic tool of chemometrics. *Chemometrics and Intelligent Laboratory Systems*, 58(2):109–130.

Yao, F., Coquery, J., and Lê Cao, K.-A. (2012). Independent Principal Component Analysis for biologically meaningful dimension reduction of large biological data sets. *BMC Bioinformatics*, 13(1):24.

Yuan, M. and Lin, Y. (2006). Model selection and estimation in regression with grouped variables. *Journal of the Royal Statistical Society: Series B (Statistical Methodology)*, 68(1):49–67.

Žitnik, M., Janjić, V., Larminie, C., Zupan, B., and Pržulj, N. (2013). Discovering disease-disease associations by fusing systems-level molecular data. *Scientific reports*, 3(1):1–9.

Zou, H. and Hastie, T. (2005). Regularization and variable selection via the elastic net. *Journal of the Royal Statistical Society: Series B (Statistical Methodology)*, 67(2):301–320.

Zou, H. and Hastie, T. (2018). *Elasticnet: Elastic-Net for Sparse Estimation and Sparse PCA*. R package version 1.1.1.

Zou, H., Hastie, T., and Tibshirani, R. (2006). Sparse principal component analysis. *Journal of Computational and Graphical Statistics*, 15(2):265–286.

Index

Note: Locators in *italics* represent figures and **bold** indicate tables in the text.